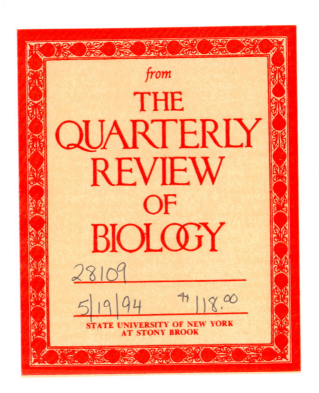

Developments in Hydrobiology 93

Series editor
H. J. Dumont

Limnology of Mountain Lakes

Edited by

J. Fott

Reprinted from Hydrobiologia, vol. 274 (1994)

Kluwer Academic Publishers
Dordrecht / Boston / London

Library of Congress Cataloging-in-Publication Data

Limnology of mountain lakes / edited by J. Fott.
p. cm. -- (Developments in hydrobiology ; v. 93)
Papers presented at a symposium held at Stará Lesná, Slovakia in July 1991.
ISBN 0-7923-2640-7 (acid free paper)
1. Aquatic organisms--Effect of water pollution on--Congresses. 2. Acid pollution of rivers, lakes, etc.--Congresses. 3. Lake ecology--Congresses. I. Fott, J. (Jan) II. Series: Developments in hydrobiology ; 93.
QH545.W3L55 1994
574.5'26322--dc20 93-44968
CIP

ISBN 0-7923-2640-7

Published by Kluwer Academic Publishers,
P.O. Box 17, 3300 AA Dordrecht, The Netherlands.

Kluwer Academic Publishers incorporates
the publishing programmes of
D. Reidel, Martinus Nijhoff, Dr W. Junk and MTP Press.

Sold and distributed in the U.S.A. and Canada
by Kluwer Academic Publishers,
101 Philip Drive, Norwell, MA 02061, U.S.A.

In all other countries, sold and distributed
by Kluwer Academic Publishers,
P.O. Box 322, 3300 AH Dordrecht, The Netherlands.

Printed on acid-free paper

Printed in Belgium

Contents

Preface

This volume contains papers presented either in oral or poster form at the international symposium Limnology of Mountain Lakes, which was held at Stará Lesná (Slovakia) between 1 and 7 July 1991. The idea of convening a conference on mountain lakes originated two years before among limnologists at the Department of Hydrobiology, Charles University, Prague. After having accomplished a decade of extensive studies on lakes in the Šumava Mountains and in the High Tatras, they were looking for a way to extend international scientific contacts in the field of mountain lake limnology and present the results of their work. The most pleasant and effective way to reach the desired goal was to organize a conference at a place close to lakes under study.

The site chosen was Stará Lesná, a small tourist center situated near the southern border of the Tatra National Park. The choice was partially influenced by the success of another conference organized in the Tatras by Vladimír Kořínek in 1989 (V. Kořínek & D.G. Frey (eds), *Biology of Cladocera [Developments in Hydrobiology 71]. Hydrobiologia* 225, 1991).

The programme included one day of excursions to mountain lakes. Altogether five trips of varying difficulty were scheduled – from a challenging hike with an overnight stay in a chalet above 2000 m, to a comfortable walk to small dystrophic forest lakes, guided by the editor of this volume. A closing bus excursion was held to the Spišská Magura and Pieniny mountains, and included rafting on the whitewater rapids of the Dunajec river.

The symposium was open to presentations from all aspects of limnology of mountain lakes. 53 participants from 13 countries presented 28 papers and 18 posters. In the two introductory lectures on lakes in the host country, Jan Fott and Evžen Stuchlík gave information concerning lakes in the Šumava (Bohemia) and in the High Tatra Mountains (Slovakia). The following oral presentations and poster sessions were contributions from the fields of physical and chemical limnology, palaeolimnology, zooplankton, phytoplankton and phytobenthos, and bacteria. Acidification, a process affecting water chemistry and biota of many mountain lakes in Europe, was dealt with in several papers, and one of the field trips was directed primarily to lakes influenced by this kind of ecological stress. A series of papers on the lakes in Šumava has highlighted different aspects of these lakes, which are in the last stage of acidification. Other geographical areas covered extensively were the Tatras and the Alps. Professor Agnes Ruttner-Kolisko, who regrettably passed away a few months afterwards, presented a contribution to ecology of "Almtümpel" in the Lunz mountain region.

A working group discussing a possible continuation of the AL:PE programme (Acidification of mountain lakes: palaeolimnology and ecology) met on one of the evenings under the chairmanship of Bente M. Wathne and Richard W. Battarbee. Two other evenings were spent in a more relaxed spirit in the folk-style "Zbojnická koliba Inn" and on the "Bears' meadow". Although no decision was made regarding a similar meeting in the future, an initiative leading to the "Limnology of Mountain Lakes II" conference would be highly appreciated.

The manuscripts offered for publication in this volume were screened and reviewed in the usual way for the acceptance of scientific papers for print. Three colleagues from overseas were forced to cancel their participance at the last moment due to insufficient funding. One of them, John G. Stockner, asked me for the possibility to publish the manuscript he had prepared for our conference, to which I agreed with pleasure. I wish to express my gratitude to David Hardekopf and Dale Osborne, who revised the English of manuscripts when necessary.

JAN FOTT
Editor

Hydrobiologia **274**, 1994.
J. Fott (ed.), Limnology of Mountain Lakes.

Organizing committee

Chairman: Jan Fott
Secretary: Evžen Stuchlík
Members: Michal Pop, Zuzana Stuchlíková, Jaroslava Dargocká, Petra Kneslová
Logistical arrangements: Jitka Daumová (Tatratour Travel Agency)

Sponsored by the Faculty of Science, Charles University, Prague. Lectures and poster exhibitions were held at the hotel HORIZONT, Stará Lesná

List of participants

ALEKSEEV V., Moscow, USSR
AMBLARD C. A., Aubière, France
BATTARBEE R. W., London, UK
BLAŽKA P., České Budějovice, Czechoslovakia
BOURDIER G., Aubière, France
BUTLER N. M., Ann Arbor, USA
CAMARERO L., Barcelona, Spain
CATALAN J., Barcelona, Spain
CRUZ-PIZARRO L., Granada, Spain
DARGOCKÁ J., Prague, Czechoslovakia
DUIGAN C. A., Aberystwyth, UK
FELIP M., Barcelona, Spain
FIKS B., Leningrad, USSR
FJELLHEIM A., Bergen, Norway
FOTT J., Prague, Czechoslovakia
GACIA E., Barcelona, Spain
GALAS J., Cracov, Poland
GEE J. H. R., Aberystwyth, UK
GRODZINSKA-JURCZAK M., Cracov, Poland
HALVORSEN G., Oslo, Norway
HOUK V., Prague, Czechoslovakia
KNESLOVÁ P., Prague, Czechoslovakia
KOČÁREK E., Prague, Czechoslovakia
KOPÁČEK J., České Budějovice, Czechoslovakia
KOT M., Zakopane, Poland
KOTOV S., Leningrad, USSR
KRAMER J. R., Hamilton, Canada
LAMI A., Verbania Pallanza, Italy
LUKAVSKÝ J., Třeboň, Czechoslovakia
MARCHETTO A., Verbania Pallanza, Italy
MENOZZI P., Parma, Italy
MORALES-BAQUERO R., Granada, Spain
MOSELLO R., Verbania Pallanza, Italy
NAUWERCK A., Mondsee, Austria
NIEDERHAUSER P., Kilchberg, Switzerland
PERNEGGER L., Innsbruck, Austria
POP M., Prague, Czechoslovakia
PRAŽÁKOVÁ M., Prague, Czechoslovakia
PRCHALOVÁ M., Kašperské Hory, Czechoslovakia
PSENNER R., Innsbruck, Austria
ŘEHÁKOVÁ Z., Prague, Czechoslovakia
ROSE N., London, UK
RUTTNER-KOLISKO A., Lunz am See, Austria
SCHMIDT R., Mondsee, Austria
ŠPORKA F., Bratislava, Czechoslovakia
STRAŠKRABOVÁ V., České Budějovice, Czechoslovakia
STUCHLÍK E., Prague, Czechoslovakia
STUCHLÍKOVÁ Z., Prague, Czechoslovakia
THIES H., Freiburg, Germany
VESELÝ J., Prague, Czechoslovakia
VRANOVSKÝ M., Bratislava, Czechoslovakia
VYHNÁLEK V., České Budějovice, Czechoslovakia
WATHNE B. M., Korsvoll, Norway

Hydrobiologia **274**: 1–7, 1994.
J. Fott (ed.), Limnology of Mountain Lakes.

Diatoms, lake acidification and the Surface Water Acidification Programme (SWAP): a review

Richard W. Battarbee
Environmental Change Research Centre, University College London, 26 Bedford Way, London WC1H 0AP, UK

Key words: lake acidification, diatoms, palaeolimnology, weighted averaging, database, quality control

Abstract

The Surface Water Acidification Programme (SWAP) was set up as collaborative research project involving scientists from Norway, Sweden and the UK. Its aim was to evaluate the factors responsible for fish decline in acid streams and lakes. A substantial sub-project was concerned with the palaeolimnological evidence for acidification and its causes. The central technique used was diatom analysis. In order to harmonise methodology between the seven diatomists from four laboratories in three countries a programme of taxonomic quality control was organised, involving slide exchanges, 'blind' counting, and regular workshops. In addition a calibration data-set of surface sediment diatoms and water chemistry from 170 lakes was constructed and archived on DISCO, the UCL diatom database. This data-set was used to generate diatom-chemistry transfer functions for pH, DOC and total Al using a weighted averaging technique. Application of the pH transfer function to sediment cores from a range of lakes demonstrated a dose-response relationship between lake sensitivity to acidification (as represented by mean Ca^{2+} values) and acid deposition (g S m^{-2} yr^{-1}), indicating the overwhelming importance of acid deposition as the cause of lake acidification.

Introduction

The Surface Water Acidification Programme (SWAP) was launched in 1983 by the Royal Society of London, the Norwegian Academy of Science and Letters and the Royal Swedish Academy of Science. It was designed primarily to evaluate the factors responsible for fishery decline in Norway, Sweden and UK. One of the main components of the research was an integrated palaeolimnology project which aimed to compare lakes and their catchments in areas of high and low acid deposition, and which sought to evaluate the various hypotheses for the cause of surface water acidification.

The results of both the full project and the palaeolimnology sub-project have been recently published (Mason, 1990; Battarbee *et al.*, 1990). This paper summarises the results of the palaeolimnology sub-project, especially those aspects concerned with the use of diatoms.

Harmonising diatom taxonomy

Because of the central role of diatom analysis in reconstructing the history of lake water chemistry, especially pH, considerable effort was committed to achieving a consistent approach to diatom taxonomy between the four main laboratories in the project. In addition, an attempt was made to standardise methodology with that used for lake acidification studies in North America by including John Kingston from the PIRLA project (Charles & Whitehead, 1986; Kingston, 1986) in the SWAP workshops.

Several aspects of taxonomy were of concern: avoiding mistakes, using agreed nomenclature, using internally consistent concepts for the splitting and amalgamation of taxa, and using agreed protocols for the inclusion and description of unknown taxa. The details of the scheme are presented by Kreiser & Battarbee (1988).

To achieve these objectives, slides containing problem or representative taxa were circulated between laboratories. Diatoms on these slides were identified and counted according to agreed procedures and the results of these 'blind' counts were compared at regular workshops. Figure 1a shows the differences in taxonomic usage before the first workshop. For many taxa, e.g. *Frustulia rhomboides* v. *saxonica*, *Navicula heimansii*, *Peronia fibula*, there were very close agreements between the laboratories. For others the differences were either due to accurate identification but use of different names, e.g. *Anomoeoneis exilis*/*A. vitrea*, inconsistent use of varietal status e.g. *A. exilis*/*A. exilis* v. *lanceolata*, or mistakes e.g. *Eunotia alpina*/*E. lunaris*. The modified diagram (Fig. 1b), constructed after the first workshop not only shows how these differences were resolved but also shows the result of using an agreed nomenclatural checklist, in this case the *Checklist of British Diatoms* (Hartley, 1986; Williams *et al.*, 1988). In some cases all four laboratories correctly identified the taxon and used the same name but this was not consistent with the checklist name. The name was consequently changed according to the checklist, e.g. *Navicula heimansii* = *N. leptostriata*, *Eunotia veneris* = *E. incisa*.

DISCO – the diatom database

During the SWAP programme diatom analyses of surface sediments and associated analyses of water chemistry and catchment descriptions were made from approximately 170 lakes sites. In addition diatom analyses of core sediments were carried out for approximately 20 sites. All diatom and environmental data were entered on the diatom database (DISCO) at University College London as described by Munro *et al.* (1990). The database uses the commercial programme ORACLE. It contains a coded version of the British Diatom Checklist and a code dictionary of all taxa used and agreed to in the SWAP programme, including unknowns. The database was used to calculate percentages, to make agreed amalgamations of taxa (e.g. *Achnanthes minutissima* and *A. microcephala* were kept separate in counts but amalgamated for pH reconstruction), and to provide a list of the more frequently occurring taxa, defined for statistical purposes as those present in at least two of the samples and having a frequency of more than 1% in at least one sample. In addition all the aggregate taxonomic groups to genus level and above (e.g. *Navicula* spp.) were deleted before combining with water chemistry data for statistical analysis and the derivation of calibration equations.

Diatom – water chemistry training set

A major advance in palaeolimnology in the last decade has been the development of quantitative techniques to relate diatom distributions to environmental gradients and the use of such relationships to generate calibration equations for environmental reconstruction (Huttunen & Meriläinen, 1986; Charles, 1985; Birks, 1987). The power of such equations depends essentially on the quality of the modern training set. In SWAP the training-set consisted of diatom, water chemistry and site information from 170 lakes covering southern Norway, southern Sweden and upland Britain. The chemical data-set is an amalgamation of sub-sets from different regions and laboratories and not all determinands were measured at all sites.

Consequently chemical data were carefully screened and harmonised before combining with

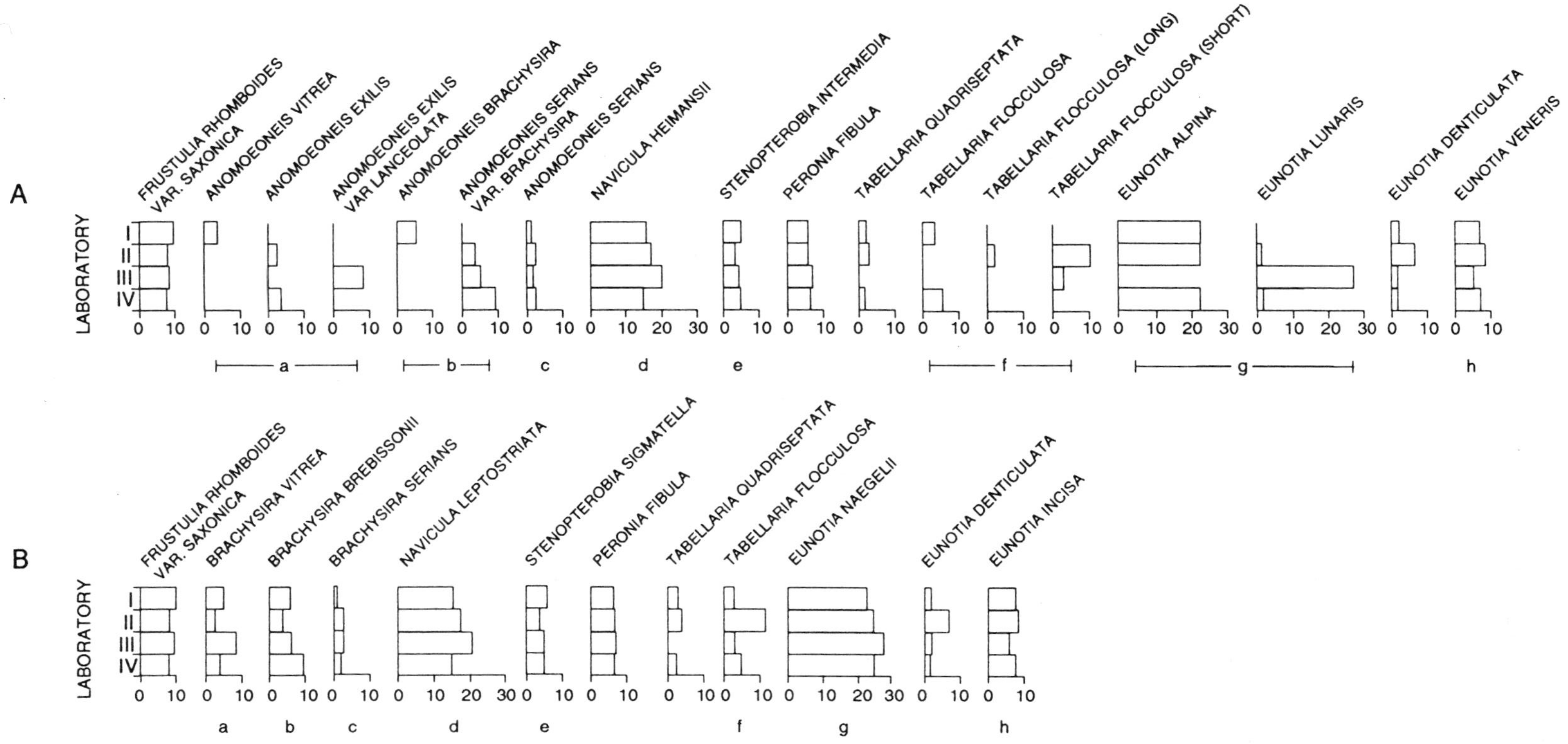

Fig. 1a. (upper) Dominant diatom taxa in slides from a sample from Lingmoor Tarn, Cumbria (provided by Dr E. Y. Haworth) showing differences between laboratories prior to quality control. The results illustrate (i) problems of nomenclature – groups a, b, c, d, e, g, and h; (ii) problems of splitting versus amalgamation – groups a and f; and (iii) the use of differing identification criteria – group g. Horizontal scale is percentage occurrence. *1b.* (lower) Dominant taxa in the Lingmoor Tarn slide after taxonomic and nomenclatural revision (from Munro *et al.*, 1990).

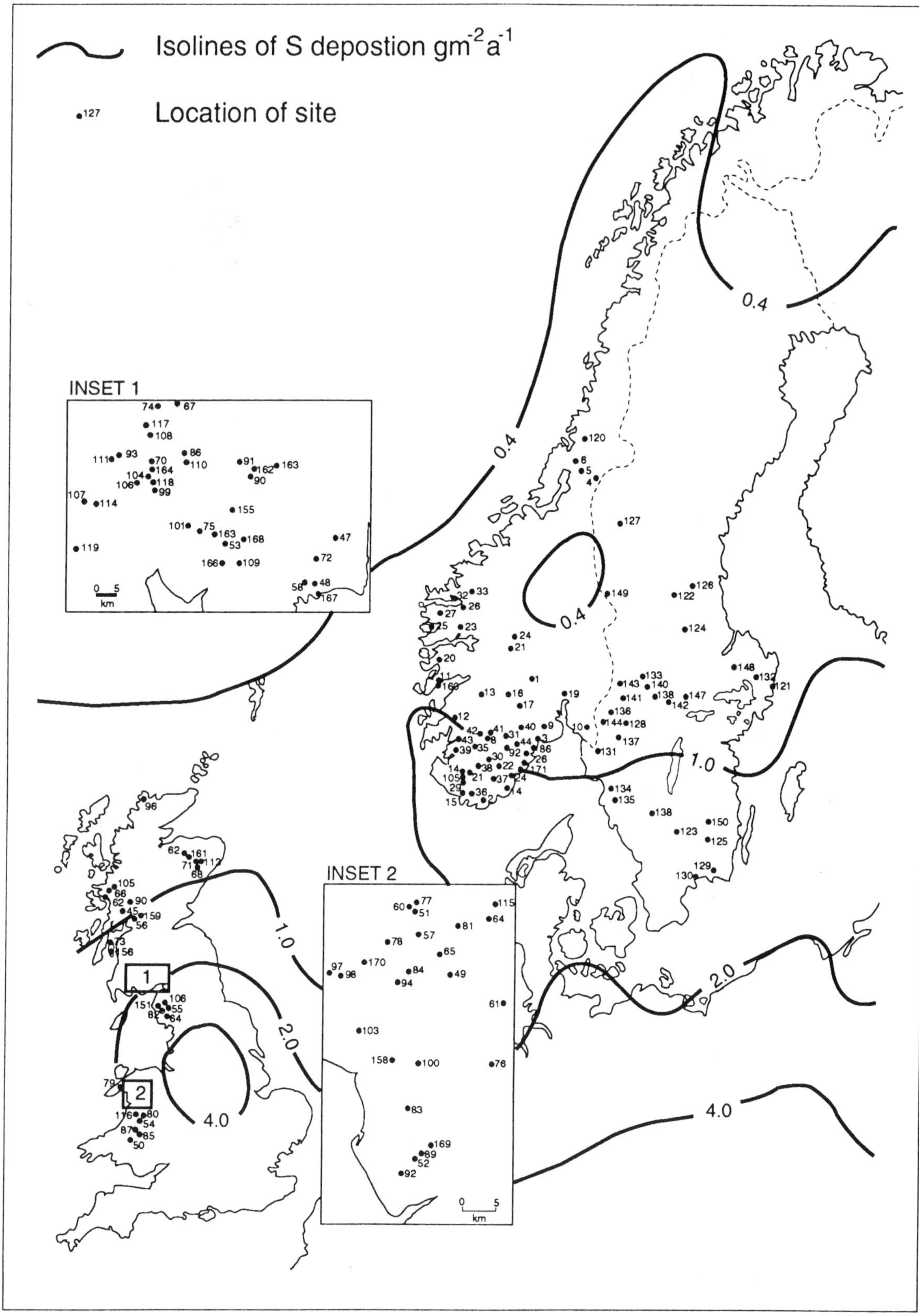

Isolines of S depostion gm-2a-1
Location of site
INSET 1
INSET 2
0.4
1.0
2.0
4.0

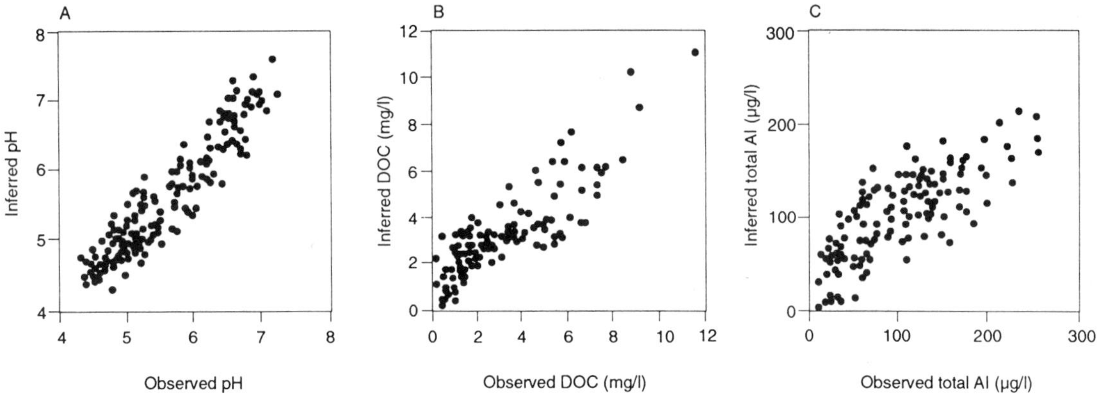

Fig. 3. Plots of inferred against observed values based on weighted averaging regression and calibration using the screened SWAP training set for (a) pH, $n = 167$; (b) DOC, $n = 123$; and (c) total Al, $n = 126$ (from Stevenson *et al.*, 1991).

the diatom data. Figure 2 shows the distribution of sites in relation to S deposition contours. A full list of sites and associated data is given in Stevenson *et al.* (1991).

Reconstructing pH, DOC and total Al

Most early pH reconstruction methods (Meriläinen, 1967; Renberg & Hellberg, 1982; Charles, 1985) used linear regression models and grouped diatoms into pH preference categories. In SWAP it was regarded that these methods were inappropriate for both biological and statistical reasons (Birks, 1987). Instead alternative procedures based on the proposals of ter Braak and van Dam (1989) involving maximum likelihood (ML) and weighted averaging (WA) were developed. A comparison between ML and WA (Birks *et al.*, 1990a) showed that WA was not only computationally quicker and simpler but gave better results in terms of its lower root-mean-square error of prediction (RMSE). Consequently the WA approach was adopted within SWAP and used for the reconstruction of pH, DOC and total Al (Birks *et al.*, 1990b). This was facilitated by the computer programs WACALIB 2.1 (Line & Birks 1990) and 3.0 (Line, Birks & ter Braak, unpub.). Figure 3 shows plots of inferred against observed values for pH, DOC and total Al and Table 1 shows the number of sites used in each training set, and the correlation (r) between the inferred and observed values. The RMSE estimated by bootstrapping (Birks 1991a) is also presented.

An empirical model for lake acidification in the UK

Although the statistical analyses of the training sets showed that pH, DOC and total Al could be reconstructed from sediment core diatom assemblages, only pH reconstruction was routinely carried out on SWAP cores. Within SWAP the technique has been used for a variety of studies: comparing recent lake acidification histories in areas of high and low acid deposition (Jones *et al.*, 1990; Berge *et al.*, 1990; Renberg *et al.*, 1990), reconstructing post-glacial trends in pH (Renberg, 1990; Atkinson & Haworth, 1990) and testing alternative hypotheses for the causes of acidification (Anderson *et al.*, 1990; Kreiser *et al.*, 1990; Birks *et al.*, 1990c).

Fig. 2. Map of the 170 sites where surface sediments for diatom analysis and samples for chemistry were obtained. Individual site details are given in Stevenson *et al.*, 1991. Sulphar deposition isolines are also shown (from Stevenson *et al.*, 1991).

Table 1. Statistical data for the three SWAP training sets after data screening: number of samples (n); correlation (r) between observed and inferred chemistry; and root-mean-square error of prediction estimated by bootstrapping (RMSE (boot)). Data from Stevenson *et al.* (1991).

Training set	*n*	*r*	RMSE (boot)
pH	167	0.933	0.320
Al ($\mu g\ l^{-1}$)	126	0.777	49.663
DOC (mg l^{-1}	123	0.837	1.580

All studies in SWAP supported the contention that recent lake acidification was primarily caused by acid deposition, and Battarbee (1990) used the diatom-based pH reconstructions for UK sites to show that there was a dose-response relationship between S deposition and lake acidification. This provisional empirical relationship is shown in Fig. 4 and indicate that recent acidification at moorland sites is not likely where the ratio of Ca^{2+} ($\mu eq\ l^{-1}$) to S deposition ($g\ m^{-2}\ y^{-1}$) is greater than 70:1.

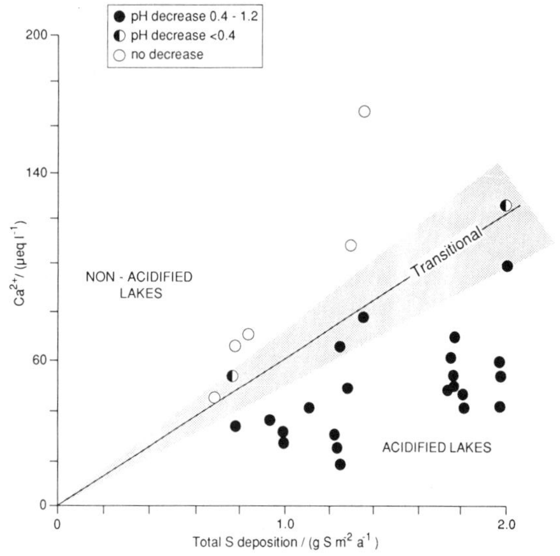

Fig. 4. Plot of lake-water calcium against total sulphur deposition for lakes in the UK where pH reconstruction has been carried out on dated sediment cores. Acidified lakes occur only when the Ca:S ratio is less than 60:1 (from Battarbee, 1990).

Coda

Although SWAP has now ended, work on the diatom/chemistry relationships is continuing. The dose-response model is currently being further developed by adding more sites and adapting it for use in the definition of critical loads (Battarbee 1989, Henriksen & Brakke 1988). The SWAP training set is being used in studies of reversibility (Allott *et al.*, 1993), and is being developed to strengthen DOC and Al reconstructions (Birks, pers. comm.) and to allow analogue matches to be made between modern and fossil assemblages (Juggins, pers. comm.).

Acknowledgements

I should like to thank all those involved in the SWAP Palaeolimnology Programme, especially Ingemar Renberg, my cocoordinator, and John Birks, the leader of the statistical subgroup, for their help and enthusiasm during the SWAP years. I am also grateful for the support we received from the SWAP Management Committee. SWAP was funded by the UK Central Electricity Generating Board and the UK National Coal Board.

References

Allott, T. E. H., R. Harriman & R. W. Battarbee, 1993. Reversibility of lake acidification at the Round Loch of Glenhead, Galloway, Scotland. Envir. Pollut. 77: 219–222.

Anderson, N. J. & T. Korsman, 1990. Land-use change and lake acidification: iron-age desettlement in Northern Sweden as a pre-industrial analogue. Phil. Trans. r. Soc., Lond. B 327: 373–376.

Atkinson, K. M. & E. Y. Haworth, 1990. Devoke Water and Loch Sionascaig: recent environmental changes and the post-glacial overview. Phil. Trans. r. Soc., Lond. B 327: 349–355.

Battarbee, R. W., 1989. The acidification of Scottish lochs and the derivation of critical sulphur loads from palaeolimnological data. Palaeoecology Research Unit Report No. 36, University College London.

Battarbee, R. W., 1990. The causes of lake acidification, with special reference to the role of acid deposition. Phil. Trans. r. Soc., Lond. B 327: 339–347.

Battarbee, R. W., J. Mason, I. Renberg & J. F. Talling, J. F. (eds), 1990. Palaeolimnology and lake acidification. The Royal Society London, 219 pp.

Berge, F., Y-W. Brodin, G. Cronberg, F. El-Daoushy, H. I. Hoeg, J. P. Nilssen, I. Renberg, B. Rippey, S. Sandoy, A. Timberlid & M. Wik, 1990. Palaeolimnological changes related to acid deposition and land-use in the catchments of two Norwegian soft-water lakes. Phil. Trans. r. Soc., Lond. B 327: 385–389.

Birks, H. J. B., 1987. Methods for pH calibration and reconstruction from palaeolimnological data: procedures, problems, potential techniques. Proceedings of the Surface Water Acidification Project (SWAP) mid-term review conference. Bergen 22–26 June 1987, London: SWAP: 370–380.

Birks, H. J. B., J. M. Line, S. Juggings, A. C. Stevenson & C. J. F. ter Braak, 1990a. Diatoms and pH reconstruction. Phil. Trans. r. Soc., Lond. B 327: 263–278.

Birks, H. J. B., S. Juggins, J. M. Line, 1990b. Lake surface-water chemistry reconstructions from palaeolimnological data. In B. J. Mason (ed.), The Surface Waters Acidification Programme. Cambridge University Press, Cambridge: 301-313.

Birks, H. J. B., F. Berge, J. P. Boyle, 1990c. A palaeoecological test of the land-use hypothesis for recent lake acidification in South-west Norway by using hill-top lakes. J. Paleolimnol. 4: 69–85.

Charles, D. F., 1985. Relationships between surface sediment diatom assemblages and lake water characteristics in Adirondack lakes. Ecology 66: 994–1011.

Charles, D. F. & D. R. Whitehead, 1986. The PIRLA project: Paleoecological investigations of recent lake acidification. Hydrobiologia 143/Dev. Hydrobiol. 37: 13–20.

Hartley, B., 1986. A check-list of the freshwater, brackish and marine diatoms of the British Isles and adjoining coastal waters. J. mar. biol. Ass. U.K. 66: 531–610.

Henriksen, A., 1988. Sulfate deposition to surface waters. Envir. Sci. Technol. 22: 8–14.

Huttunen, P. & J. Meriläinen, 1986. Applications of multivariate techniques to infer limnological conditions from diatom assemblages. In J. P. Smol, R. W. Battarbee, R. B. Davis & J. Meriläinen (eds), Diatoms and Lake Acidity. Developments in Hydrobiology 29. Dr W. Junk Publishers, Dordrecht: 201–211.

Jones, V. J., A. M. Kreiser, P. G. Appleby, Y-W. Brodin, J. Dayton, J. A. Natkanksi, N. G. Richardson, B. Rippey, S. Sandoy & R. W. Battarbee, 1990. The recent palaeolimnology of two sites with contrasting acid deposition histories. Phil. Trans. r. Soc., Lond. B 327: 397–402.

Kingston, J. C., 1986. Diatom methods. In D. F. Charles & D. R. Whitehead (eds), Paleoecological investigation of recent lake acidification methods and project description. EPRI EA-4906, Electric Power Research Institute, Palo Alto, California.

Kreiser, A. & R. W. Battarbee, 1988. Analytical quality control (AQC) in diatom analysis. In U. Miller & A-M Robertsson (eds), Proceedings of the Nordic Diatomist Meeting, Stockholm, June 10-12, 1987: 41–44.

Kreiser, A., F. Oldfield, B. Rippey, J. A. Natkanski, S. Sandoy, Y. Broding, A. C. Stevenson, S. T. Patrick & R. W. Battarbee, 1990. Afforestation and lake acidification: a comparison of four sites in Scotland. Phil. Trans. r. Soc., Lond. B 327: 377–383.

Line, J. M. & H. J. B. Birks, 1990. WACALIB version 2.1 – a computer program to reconstruct environmental variables from fossil assemblages by weighted averaging. J. Paleolimnol. 3: 170–173.

Mason, B. J. (ed.) 1990. The Surface Waters Acidification Programme. Cambridge University Press, Cambridge, 522 pp.

Meriläinen, J., 1967. The diatom flora and hydrogen ion concentration of the water. Ann. bot. fenn 4: 77–104.

Munro, M. A. R., A. M. Kreiser, R. W. Battarbee, S. Juggins, A. C. Stevenson, D. S. Anderson, N. J. Anderson, F. Berge, H. J. B. Birks, R. B. Davis, R. J. Flower, S. C. Fritz, E. Y. Haworth, V. J. Jones, J. C. Kingston & I. Renberg, 1990. Diatom quality control and data handling. Phil. Trans. r. Soc., Lond. B 327: 31–36.

Renberg, I., 1990. 12600 year perspective of the acidification of Lilla Öresjön, Southwest Sweden. Phil. Trans. r. Soc., Lond. B 327: 357–361.

Renberg, I., Y-W. Brodin, G. Cronberg, F. El-Daoushy, F. Oldfield, B. Rippey, S. Sandoy, J-E. Wallin & M. Wik, 1990. Recent acidification and biological changes in Lilla Öresjön, southwest Sweden, and the relation to atmospheric pollution and land-use history. Phil. Trans. r. Soc., Lond. B 327: 391–396.

Renberg, I. & R. W. Battarbee, 1990. The SWAP Palaeolimnology Programme: a synthesis. In B. J. Mason (ed.), The Surface Waters Acidification Programme. Cambridge University Press, Cambridge: 281–300.

Renberg, I. & T. Hellberg, 1982. The pH history of lakes in southwestern Sweden, as calculated from the subfossil diatom flora of the sediments. Ambio 11: 30–33.

Stevenson, A. C., S. Juggins, H. J. B. Birks, D. S. Anderson, N. J. Anderson, R. W. Battarbee, F. Berge, R. B. Davis, R. J. Flower, E. Y. Haworth, V. J. Jones, J. C. Kingston, A. M. Kreiser, J. M. Line, M. A. R. Munro & I. Renberg, 1991. The Surface Waters Acidification Project Palaeolimnology Programme: Modern diatom/lake-water chemistry data-set. Ensis Ltd, London, 86 pp.

ter Braak, C. J. F. & H. van Dam, 1989. Inferring pH from diatoms: a comparison of old and new calibration methods. Hydrobiologia 178: 209–223.

Williams, D. M., B. Hartley, R. Ross, M. A. R. Munro, S. Juggins & R. W. Battarbee, 1988. A coded checklist of British diatoms. ENSIS Publishing, London, 74 pp.

Hydrobiologia **274**: 9–16, 1994.
J. Fott (ed.), *Limnology of Mountain Lakes*.

Lipid storage in *Diaptomus kenai* (Copepoda; Calanoida): effects of inter- and intraspecific variation in food quality

Nancy M. Butler
Ecology Group, Department of Zoology, University of British Columbia, 2204 Main Mall, Vancouver BC V6T 1W5, Canada; Present address: Flathead Lake Biological Station, The University of Montana, 311 BioStation Lane, Polson, MT 59860, USA

Key words: Copepoda, calanoida, *Diaptomus kenai*, lipids, food, mountain lakes

Abstract

In nature, changes in nutrient levels and phytoplankton community structure can represent variation in food quality and quantity. Because zooplankton can be highly susceptible to starvation, individuals could enhance their survival, growth, and reproduction during periods of unfavorable food conditions by maintaining and utilizing energy stores. Patterns of lipid accumulation and depletion in *Diaptomus kenai*, a calanoid copepod from an oligotrophic montane lake, were monitored over a variety of food conditions. The results suggest that, while lipid storage in *D. kenai* is affected by phytoplankton community structure and abundance, total lipid stores respond more to gradual changes in food regime than to sudden changes in food supply.

Changes in lake nutrient supplies and phytoplankton community structure represent variation in the quantity and quality of food available to grazing zooplankton. These changes can occur over time scales ranging from minutes, such as microscale nutrient patches excreted by zooplankton (Lehman & Scavia, 1982), to hours or days, such as experienced during lake turnover, to seasons, such as periods between lake turnover events (Wetzel, 1983). Because zooplankton can be highly susceptible to starvation mortality (Threlkeld, 1986; Williamson *et al.*, 1985), individuals capable of accumulating and utilizing a stored energy source could potentially enhance their survival, growth, and reproduction during periods of low food availability and/or quality.

The importance of lipid stores during periods of low food availability is suggested by the direct relationship between feeding conditions and energy storage observed in marine invertebrates (Benson & Lee, 1975; Lee, 1975; Håkanson, 1984), and the enhanced survival of cladocerans with lipid stores during periods of low food supply (Tessier *et al.*, 1983; Holm & Shapiro, 1984; Cowgill *et al.*, 1985). Lipid stores may also be of importance when the food regime remains fairly constant but energetic needs fluctuate. For example, during reproduction the increased metabolic costs experienced by females may be met by utilizing stored energy reserves which are accumulated during periods of lower metabolic needs. Lipid stores of females are depleted during periods of egg production by cladocerans (Goulden *et al.*, 1982; Tessier & Goulden, 1982; Tessier *et al.*, 1983; Cowgill *et al.*, 1984; Holm & Shapiro, 1984) and marine copepods (Conover, 1962; Benson *et al.*, 1972; Lee *et al.*, 1972) while reserves of males remain unchanged.

Lipid stores are affected by food abundance and nutritional composition, resulting in changes in total lipid content as well as lipid class composition of the zooplankton. Phytoplankton reared under nitrogen-limited (Fogg, 1959; Ben-Amotz *et al.*, 1985; Parrish & Wangersky, 1987) or temperature-limited (Smith & Morris, 1980) growth conditions tend to accumulate lipids as the end products of photosynthesis. Because copepods derive triacylglycerides directly from phytoplankton lipids (Sargent & Falk-Petersen, 1988), high phytoplankton triacylglyceride levels, such as induced by periods of nutrient limitation, may result in high triacylglyceride content in zooplankton lipid stores.

Thus, both the quantity as well as the composition of lipid stores in copepods can be indicative of food conditions. This study investigates the effects of interspecific and intraspecific variation in food supply on total lipid content and lipid class composition of *Diaptomus kenai* from an oligotrophic lake.

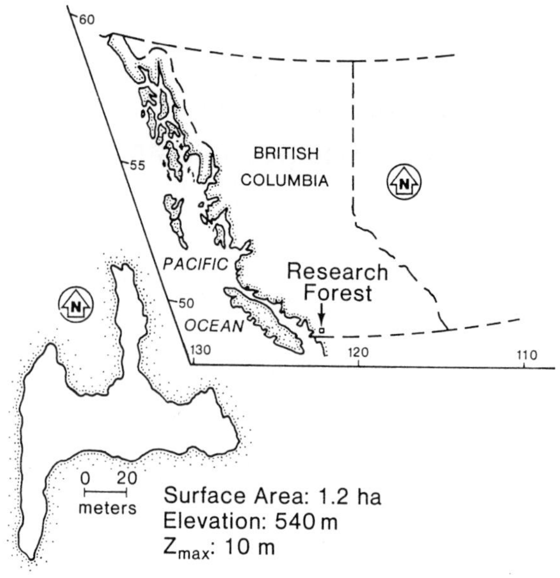

Fig. 1. Shirley Lake and location of the U.B.C. Malcolm Knapp Research Forest.

Methods

Field study

The study was conducted in Shirley Lake, a fishless, oligotrophic lake located in the University of British Columbia Malcolm Knapp Research Forest (Fig. 1). In June 1986, four big-bag enclosures constructed of 4 mil (0.1 mm) transparent polyethylene were suspended from styrofoam-supported wood frames floating at the water surface (Neill, 1978, 1981). The enclosures, measuring 1.3 m diameter by 6 m deep, were closed at the lower end to prevent contact with lake sediments and anchored in the center of the lake. Each enclosure was filled by a gasoline-powered pump with 11 000 l of lake water collected in equal volumes from 0.5 m, 2 m, and 4 m depth and filtered through a 76 μm mesh net to remove most zooplankton but not phytoplankton. Zooplankton were collected from the lake using oblique hauls of a 100 μm mesh plankton net. The collected zooplankton were pooled and distributed among the enclosures to give each a zooplankton community similar in concentration and composition to the lake community.

To produce a range of phytoplankton communities, three of the enclosures were fertilized with KH_2PO_4 and NH_4Cl at elemental N:P supply ratios of 10:1, 25:1, and 40:1. To avoid confounding desired changes in community composition with large changes in total cell density, nutrients were added at low concentrations. Orthophosphate additions (2 μg l^{-1}) were equivalent to orthophosphate levels reported for similar lakes in the Research Forest (e.g. Werring, 1986). The enclosures were fertilized weekly during the ice-free season (late March to late November) of 1986 and 1987. No nutrients were added during the winter months. The fourth enclosure received no nutrient additions. Because the purpose of fertilization was not to produce specific phytoplankton communities but simply to produce different phytoplankton communities, the nutrient manipulations themselves were not replicated.

Interspecific variations in food quality

During the period of June 1986 to October 1987, adult *Diaptomus kenai* were collected weekly with a 40 cm diameter plankton net (mesh size 100 μm) from, Shirley Lake and the experimental enclosures for lipid analysis. Immediately upon collection, copepods were isolated in dilute artificial pond water (APW: 10% dilution of the formula of Lynch *et al.* 1986), sorted according to species, sex, and, if female, reproductive condition, rinsed with distilled water, and placed in pre-weighed glass micro-culture tubes in groups of three to five individuals. Three replicates of each sex/reproductive status combination were collected, if population density and reproductive status permitted. The samples were held on dry ice for transport to the lab, where they were dried under nitrogen at 50 °C and weighed.

Lipid analysis
Total lipids were extracted and quantified as a percentage of lipid free dry mass (PLFDM) using the micro-gravimetric extraction technique of Gardner *et al.* (1985). This technique has been used successfully for analyzing small samples by other investigators (Parrish & Ackman, 1983a, b and 1985; Rao *et al.*, 1985).

For lipid class analysis, approximately 10 μl of purified extract was drawn into a 50 μl capillary pipette under a nitrogen atmosphere. The volume of the sample was measured and the tube flame-sealed under nitrogen and stored at 0 °C until the sample could be analysed by thin-layer chromatography with flame ionization detection (TLC-FID) (Parrish, 1987; Parrish *et al.*, 1988).

Initial extractions and all lipid class analyses were conducted at the National Oceanographic and Atmospheric Administration's Great Lakes Environmental Research Laboratory in Ann Arbor MI.

Laboratory study

Intraspecific variations in food quality
Selenastrum minutum was chosen for the laboratory study because it does not change size under conditions of nutrient limitation (Butler *et al.*, 1989), thereby minimizing the possibility that feeding would be affected more by changes in the physical properties of the cells than by the changes in cell chemistry.

The alga was cultured in artificial medium (Suttle & Harrison, 1988) with NH_4^+ and PO_4^- concentrations modified to 25 and 16 μM, respectively. These modifications, based on data of Elrifi and Turpin (1985), produced N-limited cells at concentrations required for the study (8–10 × 10^8 cells l^{-1}). The cultures were maintained in constant light (200 $\mu E\ m^{-2}\ s^{-1}$) at 15 °C. The cells were cultured (semi-continuous, daily dilutions) at 85% and 23% maximum potential growth rate (1.29 d^{-1} at 15 °C, Butler *et al.*, 1989) to yield high growth rate (HiGR) and low growth rate (LoGR) cells, respectively. Experiments were initiated when cell numbers and relative fluorescence (measured daily prior to dilution) varied less than 10% for a three day period, indicating steady state conditions.

Experiments were conducted in 4 l glass aquaria under constant dark at 15 °C. Six aquaria were used for each run and the entire experiment was replicated three times. All six aquaria were filled with 3.6 l of distilled water. Of these, four received a 400 ml mixture of culture medium and either LoGr or HiGR culture to yield two aquaria of each cell type at a final cell concentration of 5 × 10^5 cells ml^{-1}. Previous research indicated that feeding rate of adult female *Diaptomus kenai* on this alga is maximum at 10^4 cells ml^{-1} (Butler *et al.*, 1989). Two aquaria (the starvation treatment) received 400 ml of algal culture medium without cells. Thus, each tank was identical with respect to water chemistry (*i.e.* 10% algal culture medium) but differed in available food.

During August and September 1988, copepods were collected from Shirley Lake, transported to the lab, and sorted according to species, sex, and size. Only gravid females which could be easily classified by size as either large or small were used in the study. Because the age and condition of copepods is difficult to assess, the use of gravid females (which, by virtue of being gravid are assumed to be neither senescent nor unhealthy) reduces the possibility of obscuring the results by using copepods of indeterminant age and condition. Once sorted, the copepods were divided into groups of 10 which were then randomly assigned to 3 groups each of 100 large and 100 small *Diaptomus kenai*, one group for each food treatment. Fifty percent of the medium in each aquaria was replaced daily with fresh algal culture, medium, and distilled water to bring cell numbers to the required density. Copepods were removed from each aquarium at two day intervals in three groups of five individuals for total lipid analyses as described above. Each experiment was run until all copepods had been sampled (between 10 d and two weeks).

Statistical analyses

Total lipid content
Total lipid content (percent of lipid-free dry mass, PLFDM) of *Diatomus kenai* collected from the lake and enclosures was analyzed using two-way ANOVA of sex (male, nonreproductive female, and reproductive female) and enclosure with date as a covariate. Changes due to food quality in total lipid content (PLFDM) of large and small *D. kenai* in the laboratory were analyzed within each group by ANOVA of food treatment with time as a covariate. Regression of PLFDM on time analyzed changes in lipid stores over time in the enclosures and the laboratory. Mean PLFDM (sexes and dates combined) were compared among the enclosures using one way ANOVA with LSD (least significant difference) comparisons of the means. Data from 1986 and 1987 were analyzed separately. All data were log transformed to meet the assumptions of normality and homoscedasticity.

Lipid class composition
Lipid class composition data were compared using two-way ANOVA of enclosure and species effects with time as a covariate. Analysis was confined to three lipid groups: wax esters and triacylglycerides (the storage lipids) and polar lipids (the structural lipids). Regression of triacylglyceride composition on time analyzed temporal changes in composition within each enclosure. Lipid class composition data were available only for the 1986 samples. Data were log transformed to meet the assumptions of normality and homoscedasticity.

Results

Field Study

Total lipids
Figure 2 displays mean lipid content based on lipid-free dry mass for adult *Diaptomus kenai* (dates and sexes pooled). ANOVA comparisons of the effects of sex and enclosure on percent lipid-free dry mass (PLFDM) with date as a covariate indicate that there was a significant enclosure effect on PLFDM of *Diaptomus kenai* in 1986 ($p < 0.01$), but no significant effect in 1987 ($p = 0.10$) There was no significant covariate effect of date within either year, indicating the pattern of change in PLFDM over time was not significantly different among enclosures. Subsequent regression analysis of PLFDM on date indicated no significant time effect in either 1986 ($F = 0.504$, $p = 0.48$) or 1987 ($F = 2.231$, $p = 0.14$). One-way ANOVA with LSD indicated that, in 1986, PLFDM of *D. kenai* in the three fertilized enclosures was significantly different from that in the 'No Fertilizer' population and from that observed in the lake population ($p < 0.05$). In 1987, PLFDM of *D. kenai* in the 40 × enclosure was

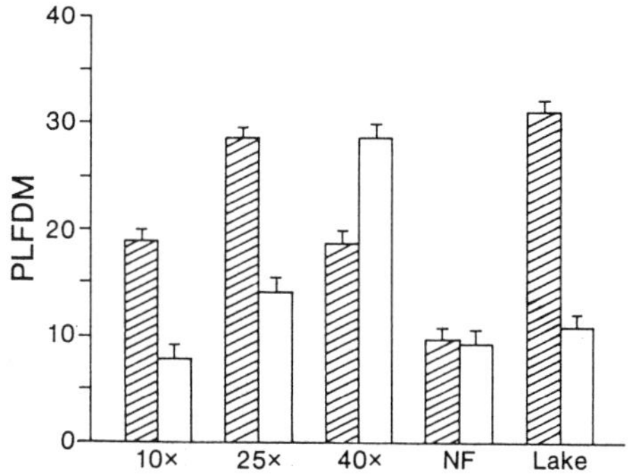

Fig. 2. Mean total lipid content (as percent lipid-free dry mass, or PLFDM), with standard error bars, of adult *Diaptomus kenai* in the lake and enclosures during 1986 (shaded columns) and 1987 (open columns).

significantly greater ($p<0.05$) than in the lake and the other enclosures, which did not differ among themselves.

Lipid class composition

The abundances of storage and structural lipids (wax esters, triacylglycerides, and polar lipids) in *D. kenai* were compared among the lake and enclosures (Fig. 3) There were significant enclosure and covariate (date) effects on triacylglyceride (storage) and polar (structural) lipid content ($p<0.05$) but not on wax ester content ($p>0.4$).

Laboratory study

Total lipid content (PLFDM) data from copepods maintained for 14 d on either a starvation, low nitrogen *Selenastrum minutum*, or high nitrogen *S. minutum* diet are presented in Fig. 4. ANOVA comparisons of total lipid content (expressed as PLFDM) indicated significant differences among the two groups of copepods tested ($p<0.001$) but no significant treatment effect and no significant interactions between species and treatment ($p>0.5$). A significant effect of the time covariate ($p<0.05$) suggests differences among the treatment and species groups in the way PLFDM changed over time. Subsequent regression analysis indicated that the observed changes, while significantly different when compared among the groups, were significant only within

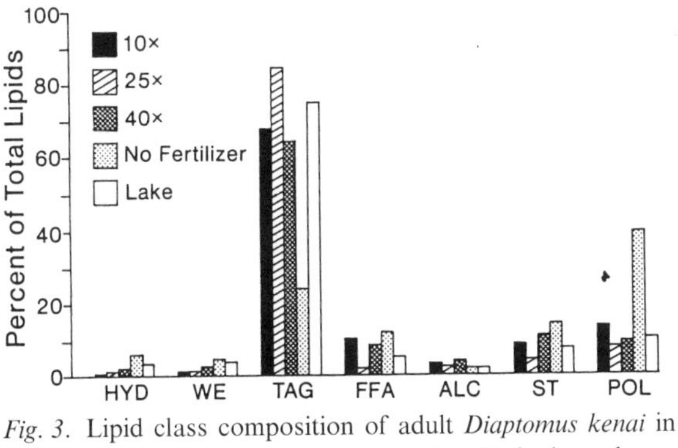

Fig. 3. Lipid class composition of adult *Diaptomus kenai* in the lake and enclosures during 1986. (HYD: hydrocarbons; WE: wax esters; TAG: triacylglycerides; FFA: free fatty acids; ALC: alcohols; POL: polar lipids)

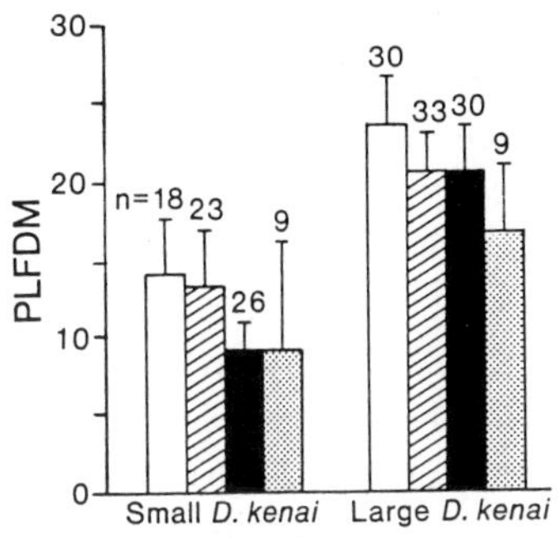

Fig. 4. Mean lipid content (as percent lipid-free dry mass, or PLFDM) of adult *Diaptomus kenai* maintained in the laboratory on no food (open columns), *Selenastrum minutum* cultured at low dilution rates (hatched columns), and *S. minutum* cultured at high dilution rates (filled columns), with standard error bars. Speckled bars indicate initial lipid content.

the large *Diaptomus kenai* feeding on the low nitrogen food ($p<0.05$).

Discussion

The observed variation in total lipid content (PLFDM) among *Diaptomus kenai* from the lake and enclosures suggests that there may indeed be variation in the food regime. While total lipid content is often a reflection of food concentration (e.g. copepods: Ikeda, 1974; Lee, 1974a; prawns: Whyte *et al.*, 1986; and oyster spat: Gallager & Mann, 1986), variation in *D. kenai* lipid content in 1986 was independent of total cell concentration in the enclosures. In the three enclosures which were fertilized (which ranged in maximum cell density from 2 to 10×10^3 cells ml^{-1}), *D. kenai* accumulated greater lipid stores than those in the nonfertilized enclosure (with a maximum cell density of 4×10^3 cells ml^{-1}), but less than those in the lake (maximum cell density 200 cells ml^{-1}). The lack of association with cell density is further supported by the fact that while cell density in the enclosures varied significantly with time, total lipid content of copepods in the enclosures was independent of time. Therefore, it would appear that lipid stores in *D. kenai* may be more a function of differences among the lake and enclosures in food quality rather than food quantity. The lack of a significant effect in 1987, however, suggests that there was no longer a substantial difference in food quality among the enclosures, despite the fact that fertilization continued through 1987. The 1987 *D. kenai* populations had very different size-class structures among the enclosures (Butler, 1990). It is possible that the different size classes of *D. kenai* (present in all the enclosures in 1986) have different food requirements, based upon such criteria as chemical composition or ingestibility, and this difference in their ability to utilize the food regime was reflected in their reproductive output, as suggested by the enclosure effect on clutch size (Butler, 1990). Thus, the population found in the enclosures in 1987 presumably represented the offspring of those copepods which were most able to utilize the 1986 phytoplankton community.

As with the majority of freshwater copepods, *Diaptomus kenai* in Shirley Lake primarily stores lipids (25 to 85% of total stores) as triacylglycerides. There was a significant enclosure effect on lipid class composition, particularly in triacylglyceride content. Triacylglyceride content of *D. kenai* in the lake varied considerably over time, with maximum levels (nearly 90% of total lipid stores) occurring in August. This peak most likely reflects decreased energy demands (it coincides with a period when small *D. kenai* are ending reproduction, but large *D. kenai* have not yet initiated reproduction) as well as natural changes in nutrient levels and phytoplankton community composition, rather than changes in absolute cell density. In the fertilized enclosures, which received a constant supply of nutrients, variation in food quality could probably be attributed to differences in species composition. In the 10 × and 40 × enclosures, triacylglyceride content of the copepods was fairly constant over time (varying less than 5%) while triacylglyceride content of copepods in the 25 × enclosure varied in a pattern similar to that in the lake. However, in the unfertilized enclosure, which did not receive a steady supply of nutrients, *D. kenai* triacylglyceride content decreased steadily over the course of the summer.

Results of the laboratory feeding study suggest that, while there were differences between the two copepod groups (small and large *Diaptomus kenai*) in total lipid content, there was no variation in lipid content attributable to differences in *Selenastrum minutum* due to nitrogen limitation. The fact that *D. kenai* is capable of discriminating between and selectively ingesting *S. minutum* cells according to degree of nitrogen limitation (Butler *et al.*, 1989) suggests that there are chemical differences between the two cell types, but these differences do not appear to be reflected in lipid content. Although lipid content of the two algal cultures was not measured, algae raised under nitrogen limited conditions are known to have a high lipid content (Reynolds, 1984; Parrish & Wangersky, 1987). However, *D. kenai* stores do not seem to reflect this intraspecific variation in food chemistry. It is possible that, because the food levels maintained during the experiment were in excess of satiation food density, the copepods offset food quality affects by altering ingestion rates, thereby maintaining lipid stores.

These results indicate that lipid content and composition of *Diaptomus kenai* can vary with feeding conditions. However, changes in lipid stores appear to be more in response to gradual (such as experienced in the field study) than to sudden (such as experienced in the laboratory study) changes in the food regime. Thus, lipid stores in *D. kenai* are more a reflection of long term or overall food conditions rather than a sensitive indicator of current feeding conditions.

Acknowledgements

This research was supported by a Natural Science and Engineering Research Council of Canada Operating Grant to William E. Neill and by a University of British Columbia Graduate fellowship. Earlier versions of this manuscript were substantially improved by the comments of C. J. Walters, P. J. Harrison, A. G. Lewis, T. H-. Carefoot, and especially W. E. Neill. Use of the Iatroscan was made possible through the generosity of Wayne S. Gardner at the Great Lakes Environmental Research Laboratory. Finally, I am forever indebted to Jim Lowden and the *Feng Shui* for helping to keep things in perspective and to Posy and Poppy, without whose help this research would have been done in half the time.

References

Ben-Amotz, A., T. G. Tornabene & W. H. Thomas, 1985. Chemical profile of selected species of microalgae with emphasis on lipids. J. Phycol. 21: 72–81.

Benson, A. A. & R. F. Lee, 1975. The role of wax in oceanic food chains. Sci. Amer. 232: 76–86.

Benson, A. A., R. F. Lee & J. C. Nevenzel, 1972. Wax esters: major marine metabolic energy sources. In J. Ganguly & R. M. S. Smellic (eds), Current Trends in the Biochemistry of Lipids. Academic Press, London: 175–187.

Butler, N. M., 1990. Responses of *Diaptomus* spp. from an oligotrophic lake to variations in food quality. Dissertation, University of British Columbia, 162 pp.

Butler, N. M., C. A. Suttle & W. E. Neill, 1989. Discrimination by freshwater zooplankton between single algal cells differing in nutritional status. Oecologia 78: 368–372.

Conover, R. J., 1962. Metabolism and growth in *Calanus hyperboreus* in relation to its life cycle. Rapp. Proc.-Verb. Cons. Int. Explor. Mer. 153: 190–197.

Cowgill, U. M., D. M. Williams & J. B. Esquivel, 1984. Effects of maternal nutrition on fat content and longevity of neonates of *Daphnia magna*. J. Crust. Biol. 4: 173–190.

Cowgill, U. M., K. I. Keating & I. T. Takahashi, 1985. Fecundity and longevity of *Ceriodaphnia dubia*/*affinis* in relation to diet at two different temperatures. J. Crust. Biol. 5: 420–429.

Elrifi, I. R. & D. H. Turpin, 1985. Steady-state luxury consumption and the concept of optimum nutrient ratios: a study with phosphate and nitrate limited *Selenastrum minutum* (Chlorophyta). J. Phycol. 21: 592–602.

Fogg, G. E., 1959. Nitrogen nutrition and metabolic patterns in algae. Symp. Soc. exp. Biol. 13: 106–125.

Gallager, S. M. & R. Mann, 1986. Individual variability in lipid content of bivalve larvae quantified histochemically by absorption photometry. J. Plankton Res. 8: 927–937.

Gardner, W. S., W. A. Frez, E. A. Cichocki & C. C. Parrish, 1985. Micromethod for lipids in aquatic invertebrates. Limnol. Oceanogr. 30: 1099–1105.

Goulden, C. E., L. Henry & A. J. Tessier, 1982. Body size, energy reserves, and competitive ability in three species of Cladocera. Ecology 63: 1780–1789.

Håkanson, J. L., 1984. The long and short term feeding condition in field-caught *Calanus pacificus*, as determined from the lipid content. Limnol. Oceanogr. 29: 794–804.

Holm, N. P. & J. Shapiro, 1984. An examination of lipid reserves and the nutritional status of *Daphnia pulex* fed *Aphanizomenon flos-aquae*. Limnol. Oceanogr. 29:1137–1140.

Ikeda, T., 1974. Nutritional ecology of marine zooplankton. Mem. Fac. Fish. Hokkaido Univ. 22: 1–97.

Lee, R. F., 1974. Lipid composition of the copepod *Calanus hypertoreas* from the Arctic Ocean. Changes with depth and season. Mar. Biol. 26: 313–318.

Lee, R. F., 1975. Lipids of Arctic zooplankton. Comp. Biochem. Physiol. 51B: 263–266.

Lee, R. F., J. Hirota, J. C. Nevenzel, R. Sauerheber, A. A-. Benson & A. Lewis, 1972. Lipids in the marine environment. Calif. Mar. Res. Comm., CalCOFI Rep. 16: 95–102.

Lehman, J. T. & D. Scavia, 1982. Microscale patchiness of nutrients in plankton communities. Science 216: 729–730.

Lynch, M., L. J. Weider & W. Lampert, 1986. Measurement of the carbon balance in *Daphnia*. Limnol. Oceanogr. 31: 17–33.

Neill, W. E., 1978. Experimental studies on factors limiting colonization by *Daphnia pulex* Leydig of coastal montane lakes in British Columbia. Can. J. Zool. 56: 2498–2507.

Neill, W. E., 1981. Impact of *Chaoborus* predation upon the structure and dynamics of a crustacean zooplankton community. Oecologia (Berl.) 48: 164–177.

Parrish, C. C., 1987. Separation of aquatic lipid classes by Chromarod thin-layer chromatography with measurement by Iatroscan flame ionization detection. Can. J. Fish. aquat. Sci. 44: 722–731.

Parrish, C. C. & R. G. Ackman, 1983a. Chromarod separations for the analysis of marine lipid classes by Iatroscan thin-layer chromatography-flame ionization detection. J. Chromatogr. 262: 103–112.

Parrish, C. C. & R. G. Ackman, 1983b. The effect of developing solvents on lipid class quantification in chromarod thin layer chromatography/flame ionization detection. Lipids 18: 563–565.

Parrish, C. C. & R. G. Ackman, 1985. Calibration of the Iatroscan-Chromarod system for marine lipid class analyses. Lipids 20: 521–530.

Parrish, C. C. & P. J. Wangersky, 1987. Particulate and dissolved lipid classes in cultures of *Phaeodactylum tricornutum* grown in cage culture turbidostats with a range of nitrogen supply rates. Mar. Ecol. Prog. Ser. 35: 119–128.

Parrish, C. C., X. Zhou & L. R. Herche, 1988. Flame ionization and flame thermionic detection of carbon and nitrogen in aquatic lipid and humic-type classes with an Iatroscan Mark IV. J. Chromatogr. 435: 350–356.

Rao, G. A., D. E. Riley & E. C. Larkin, 1985. Comparison of the Thin Layer Chromatography/Flame Ionization Detection system with other methods for the quantitative analysis of liver lipid contents in alcohol-fed rats and controls. Lipids 20: 531–535.

Reynolds, C. S., 1984. The Ecology of Freshwater Phytoplankton. Cambridge Univ. Press, 384 pp.

Sargent, J. R. & S. Falk-Petersen, 1988. The lipid biochemistry of calanoid copepods. Hydrobiologia 167/168/Dev. Hydrobiol. 47: 101–114.

Smith, A. E. & I. Morris, 1980. Synthesis of lipid during photosynthesis by phytoplankton of the Southern Ocean. Science 207: 197–198.

Suttle, C. A. & P. J. Harrison, 1988. Ammonium and phosphate uptake rates, N:P supply ratios, and evidence for N and P limitation in some oligotrophic lakes. Limnol. Oceanogr. 33: 186–202.

Tessier, A. J. & C. E. Goulden, 1982. Estimating food limitations in cladoceran populations. Limnol. Oceanogr. 27: 707–717.

Tessier, A. J., L. L. Henry, C. E. Goulden & M. W. Durand, 1983. Starvation in *Daphnia*: Energy reserves and reproductive allocation. Limnol. Oceanogr. 28: 667–676.

Threlkeld, S. T., 1986. Differential temperature sensitivity of two cladoceran species to resource variation during a blue-green algal bloom. Can. J. Zool. 64: 1739–1744.

Werring, J., 1986. Fertilization of an oligotrophic coastal montane lake using frequent doses of high N:P ratio fertilizers: Effects on phytoplankton and zooplankton community structure. M. Sci. Thesis, U.B.C, 295 pp.

Wetzel, R. G., 1983. Limnology. Saunders College Publishing, Philadelphia, 860 pp.

Whyte, J. N. C., J. R. Englar, B. L. Carswell & K. E. Medic, 1986. Influence of starvation and subsequent feeding on body composition and energy reserves in the prawn *Pandalus platyceros*. Can. J. Fish. aquat. Sci. 43: 1142–1148.

Williamson, C. E., N. M. Butler & L. Forcina, 1985. Food limitations in naupliar and adult *Diaptomus pallidus*. Limnol. Oceanogr. 30: 1283–1290.

Hydrobiologia **274**: 17–27, 1994.
J. Fott (ed.), Limnology of Mountain Lakes.

Nitrogen in the Pyrenean lakes (Spain)

J. Catalan[1,2], L. Camarero[1,2], E. Gacia[1,2], E. Ballesteros[2,3] & M. Felip[1,2]
[1]*Department of Ecology, University of Barcelona, Diagonal 645, 08028 Barcelona, Spain;* [2]*Institute of High-Mountain Research, University of Barcelona;* [3]*Centre for Advanced Studies, CSIC, Blanes*

Key words: nitrate, ammonium, nitrite, high-mountain lakes, aquatic macrophytes

Abstract

Lakes in the Pyrenees show a broad variability in nitrogen content and in the distribution of its different oxidation forms, which has no direct relation with any single physiographical, chemical or trophic feature of the lakes. Concentration of bound nitrogen in rain is low compared with other European mountains, but the annual load lies in the middle range. Seasonal and local variation in the composition of rainwater mainly depends on the geographical origin of the storms. Catchment and in-lake processes introduce further variability: NH_4^+, which is at similar concentration to NO_3^- in the rain, is quickly oxidized or adsorbed in the catchment; aquatic macrophytes can either reduce mean NO_3^- concentration in lake water (*Ranunculo-Potametum*) or greatly increase it in sediment pore water (*Isoetes*); NO_2^- depends on pH; decomposition of particulate nitrogen in sediments changes with depth; lakes act as traps for dissolved inorganic nitrogen; changes in dissolved organic nitrogen suggest high microbial activities even in cold waters; melting period introduces most of the seasonal variability.

Introduction

High-mountain lakes are good sensors of regional acidification and aerial pollution as they are pristine areas, far away from local sources of acidification and pollution. Although human emission of sulphur oxides has stabilized and is even decreasing in many European countries, release of nitrogen oxides to the atmosphere is increasing (Mosello *et al.*, 1985) and will continue to do so in the near future. Therefore, the relative importance of nitrogen compounds as acidifying agents is also rising and more attention is being paid to their significance (Schuurkes & Mosello, 1988).

In order to take advantage of mountain lakes as environmental sensors, it is necessary to understand their ecological dynamics and, in this particular case, the biogeochemical nitrogen cycle. This cycle is mainly driven by microbiological processes, but other biological and physical processes should not be neglected. In this paper we present an overview of current studies on nitrogen in the lakes in the Pyrenees, emphasizing those aspects that are more peculiar to these lakes and which must be taken into account in a comprehensive view of the nitrogen cycle in mountain lakes.

Study area and methods

The location of the sampled lakes and the precipitation sampling stations on the Central and Eastern Pyrenees (NE Spain) is shown in Fig. 1.

Bulk precipitation samples were collected from August 1987 to August 1988 at four stations:

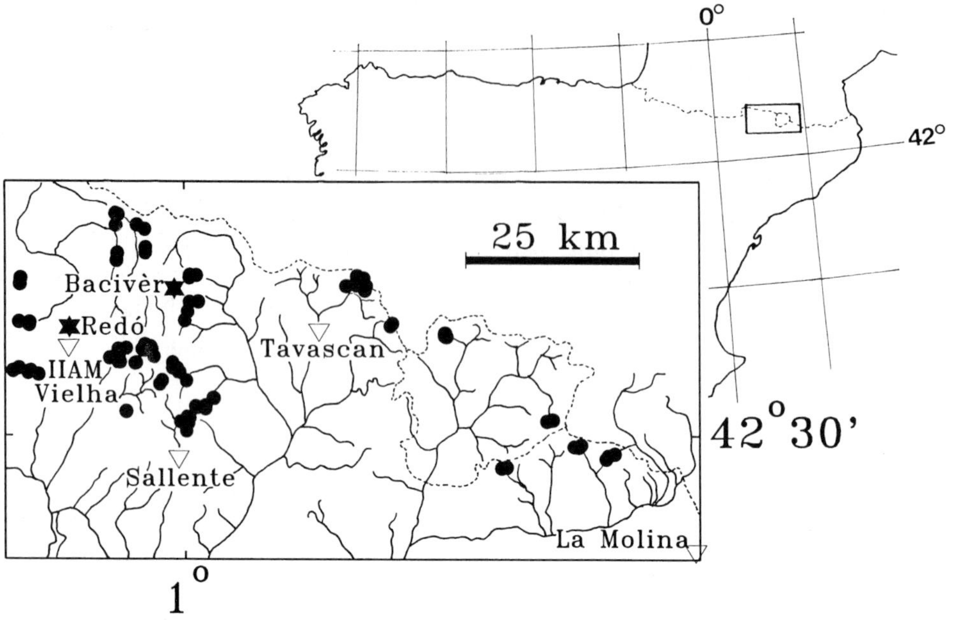

Fig. 1. Map of the area of the Pyrenees covered by this study. Circles represent the lakes samples during the 1987 survey. Inverted triangles are bulk precipitation sampling stations. Stars show lakes where seasonal sampling was carried out.

La Molina (42° 37′ 30″ N, 1° 57′ 14″ E; altitude 1704 m), Tavascan (42° 38′ 55″ N, 1° 15′ 14″ E; altitude 1125 m), Sallente (42° 28′ 00″ N, 0° 59′ 24″ E; altitude 1420 m), and Vielha (42° 37′ 20″ N, 0° 46′ 12″ E; altitude 1600 m). Details of sampling can be found in Camarero & Catalan (1993).

A total number of 102 lakes were sampled from July to September 1987. Samples were taken in the deepest point of the lake, at a depth of half the Secchi disk depth, by means of a Ruttner bottle. Details of location, morphology, vegetation and geology are given in Catalan *et al.* (1993). Granodiorite bedrocks predominate (80%), vegetation is usually herbaceous, and soils are mainly of a ranker type, usually poorly developed.

Seasonal studies were carried out in a deep cirque lake (Lake Redó, 42° 38′ 34″ N, 0° 46′ 13″ E; altitude 2240 m; maximum depth 73 m; surface area 24 ha) and a shallow valley lake (Lake Bacivèr, 42° 41′ 46″ N, 0° 59′ 1″ E; altitude 2100 m; maximum depth 6 m; surface area 2.7 ha). Lake Redó (LR) has a high residence time of water (> 4 years) and is located in a small catchment with little vegetation (Catalan, 1988); Lake Bacivèr (LB) has a mean residence time of a few days and its catchment is larger and may be divided in two basins of approximately equal areas which drain into the lake through two main inlets in the Northern and Eastern shores (Ballesteros *et al.*, 1989). Several lakes are located at higher altitude in the Northern catchment. There are 6 species of aquatic macrophytes which can be assembled in three communities (Ballesteros *et al.*, 1989): a) the *Isoetes lacustris* community, almost monospecific, is located 0.5 m–2.3 m deep in the flowless parts of the lake; b) the *Sparganium angustifolium* community is located near the inlets and outlets of the lake in areas shallower than 1 m. It is composed of an upper stratum of *Sparganium* leaves and a basal stratum of *Subularia aquatica*, *Isoetes setacea*, and *Eleocharis acicularis*, but most of the biomass is of *Sparganium*; c) finally, *Nitella gracilis* covers the deepest part of the lake below 2.5 m.

LR was sampled at eleven different depths in the deepest part, twice a month for more than a year (May 1984 to August 1985). LB was sampled monthly from October 1987 to October 1988, and water samples were taken at 2 m depth intervals at the deepest point, at two macrophyte beds and from the inlets and outlet. Sediment samples were taken in two of the main macrophyte beds using SCUBA equipment (13 July, 1987). Three cores of 12.5 cm diameter and 20 cm length were taken at 0.8 m, 1.8 m, and 2.3 m depth in the *Isoetes* community and at 0.6 m depth in the *Sparganium* community. Settling matter was determined by staking sediment traps (consisting of PVC cylinders of 5.3 cm diameter and 30 cm length) 15 cm into the sediment at 0.8, 1.8, 2.3 and 4 m depth, which were left from early spring to late autumn of 1990. *Isoetes* samples for nutrient content in tissues were taken once a month from October 1988 to October 1989.

NH_4^+ was determined according to Solórzano (1969). NO_3^- was determined by light absorbtion at 220 nm wavelength and corrected for organic matter interference (Slanina *et al.*, 1976). NO_2^-, NO_3^- from LR and dissolved organic nitrogen (DON) were determined following Grasshoff *et al.* (1983). Particulate nitrogen (PN) was determined by means of a CNS Carlo Erba 1500 Analyzer. *Isoetes* leaves were dried at 70 °C to constant weight, ground and N content was determined as PN above. N requirements for annual macrophyte production were estimated from data on annual primary production of the plants (Gacia & Ballesteros, 1991; Gacia, 1992).

Sediment cores were sliced in 2.5 cm subsamples. Fresh weight; dry weight (after drying at 105 °C to constant weight), water content and volume of the sediment were determined. Interstitial water extraction was performed according to Levat *et al.* (1990).

Results and discussion

Atmospheric input

Contribution from the atmosphere is the main source of nitrogen compounds in mountain catchments. Mean values in the Pyrenees were 17 $\mu mol\ l^{-1}$ NO_3^- and 21 $\mu mol\ l^{-1}$ NH_4^+, which lie in the lower range of the values recorded in other mountain locations at similar altitude in Central Europe (NO_3^- 17–43 $\mu mol\ l^{-1}$, NH_4^+ 21–59 $\mu mol\ l^{-1}$) (Keller *et al.*, 1987; Meesemburg, 1987; Mosello *et al.*, 1987, 1988; Psenner, 1987). NH_4^+ and NO_3^- concentrations were significantly correlated ($p<0.05$), as well as SO_4^{2+} and NH_4^+ (Camarero & Catalan, 1993). Given that most of these ions are suspended in the atmosphere in the form of $(NH_4)_2SO_4$ and NH_4NO_3 aerosols (Cadle *et al.*, 1985), this relationship could be expected. NO_2^- was always low (mean 0.16 $\mu mol\ l^{-1}$), and its distribution was not correlated with any other nitrogen compound.

Comparing different sites, Vielha, the Western station, presented higher values for all nitrogen compounds. The values were 70% higher for NH_4^+ and 37% higher for NO_3^-, which is due to a higher proportion of precipitation from the North and Northwest in Vielha than in the other sites, because of its location and orography. In fact, rains from the North and Northwest have higher contents of NO_3^- and NH_4^+ and are also more acidic (Camarero & Catalen, 1993).

Despite the relatively low concentration, total deposition was appreciable, depending as it does on the volume of precipitation. Total inorganic nitrogen (TIN) loadings ranged from 4.0 to 10.3 kg ha^{-1} yr^{-1}, comparable to deposition in moderately polluted forested areas in South Sweden and other areas on the European continent (Grennfelt & Hulberg, 1986).

There was a seasonal pattern in the deposition. Maximum mean seasonal concentrations for both NH_4^+ (26.0 $\mu mol\ l^{-1}$) and NO_3^- (22.7 $\mu mol\ l^{-1}$) were reached during summer, and minimum values during autumn (NH_4^+, 9.2 $\mu mol\ l^{-1}$; NO_3^-, 8.8 $\mu mol\ l^{-1}$). This has already been reported by several authors (Buishand *et al.*, 1988; Young *et al.*, 1988). Spring NH_4^+ values were similar to summer but NO_3^- was lower (12.8 $\mu mol\ l^{-1}$). Apart from possible local causes, it is relevant that a change in the geostrophic circulation of the air masses during autumn that favours storms coming from the South; these rainstorms are poorer in TIN than Northern and Northwestern ones.

Inter-lake variability

Figure 2 shows the concentration distribution of NO_3^-, NO_2^- and NH_4^+ in 102 lakes in the Pyrenees. Mean values of NO_3^- and, especially, NH_4^+ are significantly lower than in the rain.

Variability of NO_3^- does not show any significant correlation, either with other chemical parameters sampled at the same time (alkalinity, pH, conductivity, Ca^{2+}, Mg^{2+}, Na^+, K^+, TP, SO_4^{2-}, Cl^-, NO_2^-, NH_4^+), or with morphological or physiographical parameters (lake area, catchment area, ratio between lake and catchment areas, depth, altitude, catchment vegetation, bedrock) (Catalan *et al.*, 1993).

Comparing NO_3^- values from our survey and values from other surveys carried out from 1974 to 1976 in some of the lakes during the same period of the year (Campàs & Vilaseca, unpublished) (Fig. 3), it appears that NO_3^- has increased. In these data, it is interesting to notice that Lake Malniu is still showing low values similar to those reported 12 years ago, suggesting a persistence in the mechanism determining these values.

It has been suggested that the differences in the distribution of NO_3^- in the lakes of different mountain chains (Psenner, 1989) may be due, in addition to the main effect of the different composition of precipitation, to the presence of aquatic macrophytes in the lakes of some chains and their absence in others (Psenner, pers. comm.). In the Pyrenees, two main types of macrophyte communities are found: *Isoeto-Sparganietum* and *Ranunculo-Potametum* (Ballesteros & Gacia, 1991). Lakes with the *Ranunculo-Potametum* community had a significantly lower NO_3^- content (ANOVA, $p<0.013$) than lakes without any macrophytes (Fig. 4). Lakes with *Isoeto-*

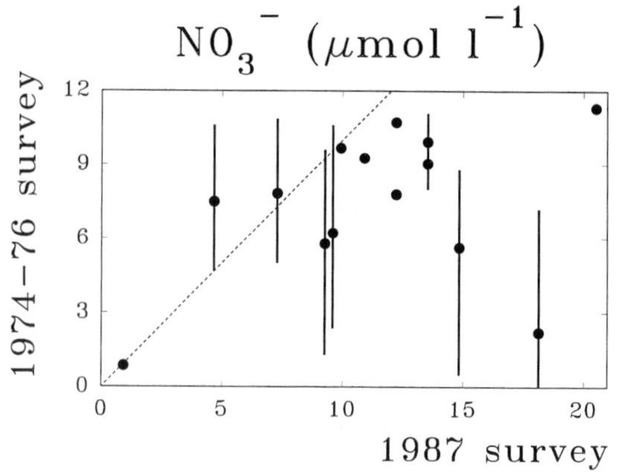

Fig. 3. NO_3^- concentration in some lakes during 1987 summer survey versus NO_3^- in the same lakes during several summer surveys between 1974–76.

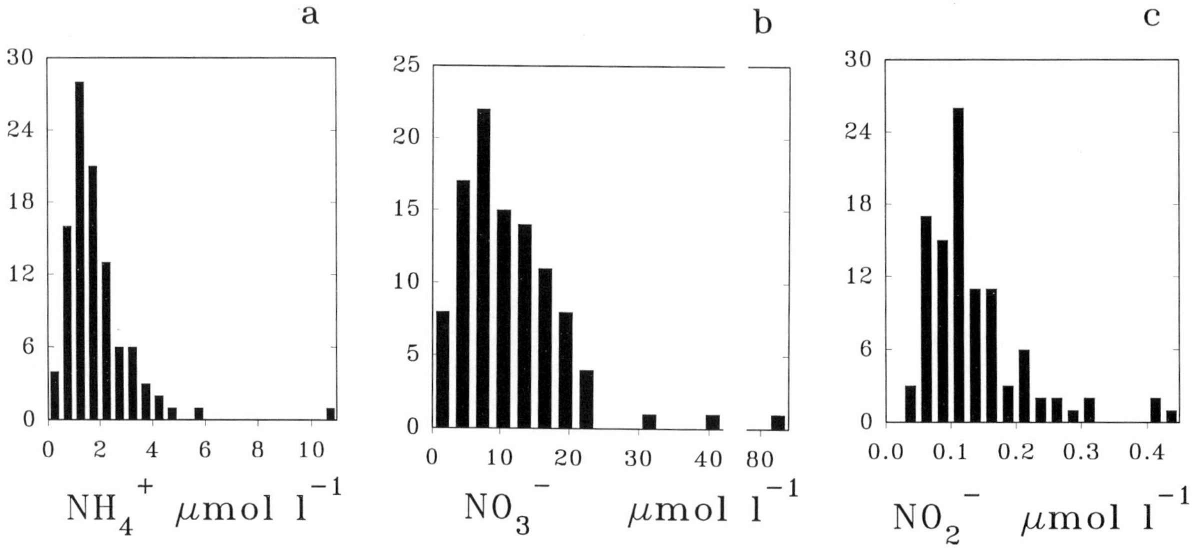

Fig. 2. Histogram distributions of NO_3^-, NO_2^- and NH_4^+ in the lakes of the Pyrenees during the 1987 survey.

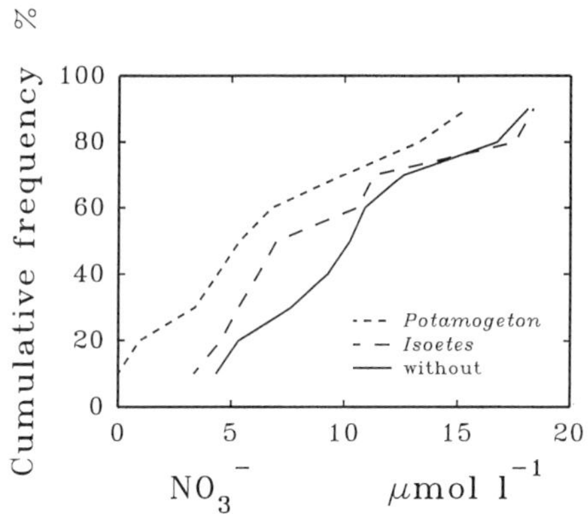

Fig. 4. Cumulative frequency distribution of NO_3^- in 3 sets of lakes of similar morphometric features and characterized by possessing *Ranunculo-Potametum* macrophyte communities, *Isoetes* communities or no aquatic macrophytes, respectively.

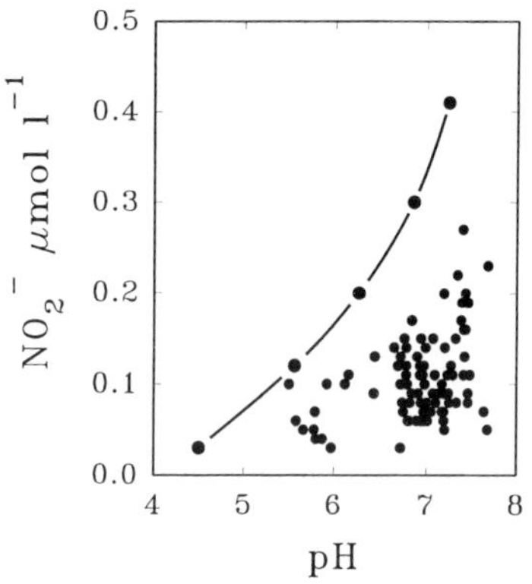

Fig. 5. Relationship between NO_2^- and pH in lakes without cattle influence.

Sparganietum did not show a globally significant departure, but the frequency distribution was appreciably different for the range below 10 μmol l^{-1} (Fig. 4). These differences do not imply that there is a direct causal relationship between macrophytes and NO_3^- levels, but it does suggest that the quantitative influence of macrophytes in the nitrogen cycle in mountain lakes should be investigated.

Nearly 40% of the lakes had NH_4^+ values <1 μmol l^{-1}, which implies that algae also uptake NO_3^- as nitrogen source (McCarthy, 1980). However, the effect on NO_3^- concentration was negligible because of the high levels of NO_3^- in relation to the productivity requirements. In fact, values of NO_3^- were slightly higher in lakes with low NH_4^+.

For the whole sample set, NH_4^+ and NO_2^- significantly correlated with Cl^-, TP and K^+ ($p<0.001$) owing to high values of all these compounds in some small eutrophied lakes because of the influence of cattle (Catalan *et al.*, 1993). Selecting lakes with Cl^- $p<18$ μmol l^{-1} (the mean rain value), this correlation disappears, and NO_2^- then strongly correlates with pH, Ca^{2+} and alkalinity ($p<0.001$). pH values established an upper limit for NO_2^- levels (Fig. 5), which might be related to the dependence of nitrification on pH. In a given lake the level of NO_2^- is stable within a narrow range, suggesting that the involved microbial processes reach a steady state.

Seasonal changes within lakes

Although LR and LB are quite different in morphology, they showed common trends in nitrogen seasonality. Particularly important in this variability was the melting period. Nitrogen concentration in the snow varies greatly (Catalan, 1989). Every snowfall adds a layer of different composition to the snowpack. During periods of high radiation, the snowpack undergoes partial melting that results in a preferential downward elution of ions (Morris & Thomas, 1985; Brimblecombe *et al.*, 1987) which causes a higher nitrogen content in the slush layers (Catalan, 1989). In LB concentrations in the slush layer were 52% higher than in the surface snow for NH_4^+, 100% higher for NO_2^-, and 302% higher for NO_3^-.

During the early snowmelt, runoff solute concentration is high; about 70–80% of solutes are carried out in a volume of 20–30% of the water stored in the snowpack (Abrahams *et al.*, 1985). Then, during late melt, runoff increases and the remaining salts yield dilute meltwater. In our study, NH_4^+ increased in LR and LB at the end of the ice-covered period (Fig. 6, 7). The peak appeared 15–30 days before the complete melting of the cover (in LR later in 1984) and was followed by a sharp decline caused by inflow (Figs 6, 7). In LB, an increase from the bottom during the ice-covered period until mid May was followed by a further increase due to the water entering the lake. In this lake, outflow concentra-

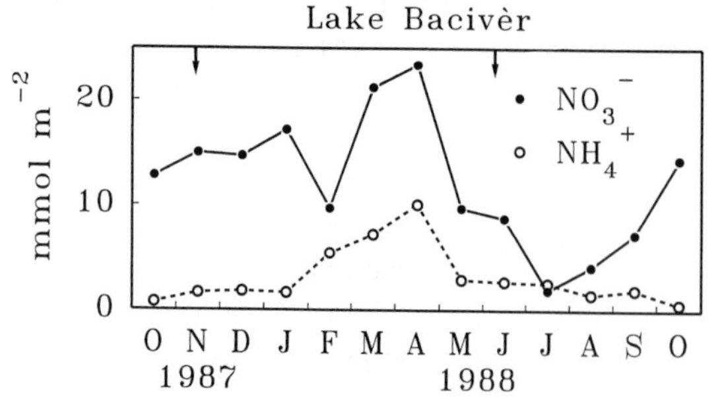

Fig. 6. Lake-wide changes in NH_4^+ and NO_3^- in Lake Bacivèr. Arrows indicate the ice cover period.

tion of NH_4^+ matched the inflow and there was little vertical differentiation during the ice-free period. In LR, during summer, NH_4^+ was higher in the epilimnion, and showed a minimum in the chlorophyll maximum below the thermocline. During autumn overturn and the early ice-covered period, the most productive phases of the year (Catalan & Camarero, 1991), NH_4^+ decreased below 0.5 μmol l^{-1}.

NO_3^- showed a peak at the end of the ice period in LB, and a minimum during July. During the ice-covered period, inflow determined higher values just under the ice. In LR, there were two large increases of NO_3^- during the sampled period: in 1984's autumn overturn and in 1985's thaw period. However, during the 1984 melting period, there was no NO_3^- peak. The difference between the two years was that during 1984 the melting of the ice cover took more time, until the end of June, and, then, the length of the spring isothermy was shorter, the mixing upon the fine bottom sediments probably being weaker. During summer stratification, both years presented a NO_3^- minimum immediately below the thermocline just above the chlorophyll maximum. NO_3^- decline in the whole column during the early ice-covered period was matched by a high productivity phase at this point, about 0.2 g C m^{-2} d^{-1} (Catalan, 1992).

In LB, the Northern inlet received water that had previously circulated through two lakes, while the Eastern inlet received water from a rocky catchment without lakes. NO_3^- was lower in the Northern inlet (Fig. 8), which might indicate a

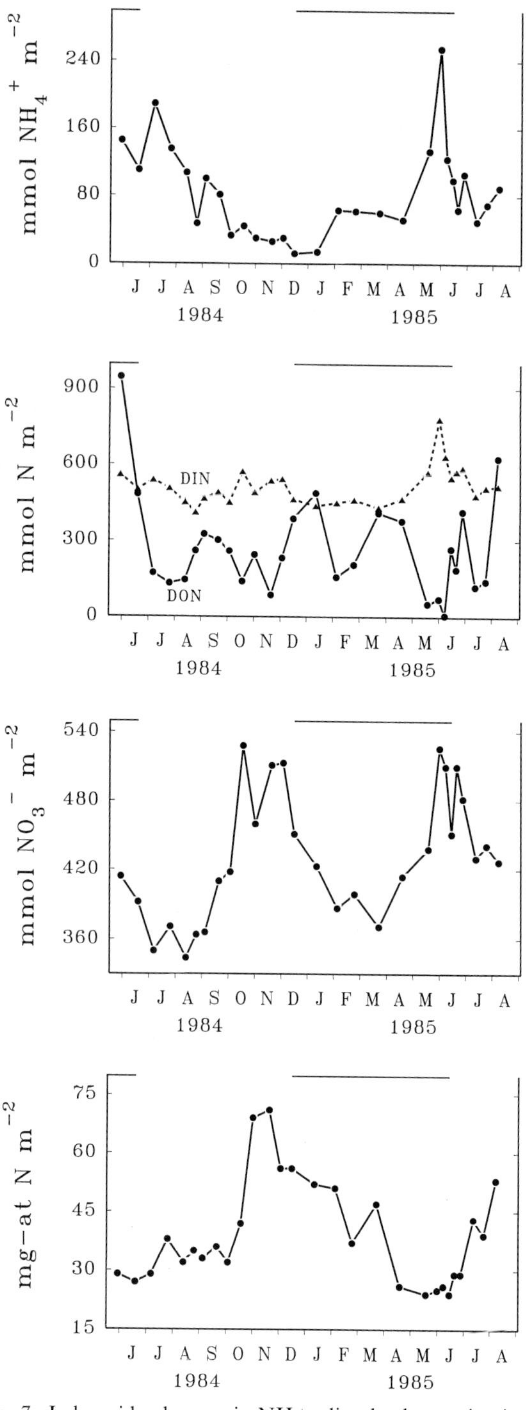

Fig. 7. Lake-wide changes in NH_4^+, dissolved organic nitrogen, dissolved inorganic nitrogen, NO_3^- and particulate nitrogen in Lake Redó. The horizontal line indicate the ice cover period.

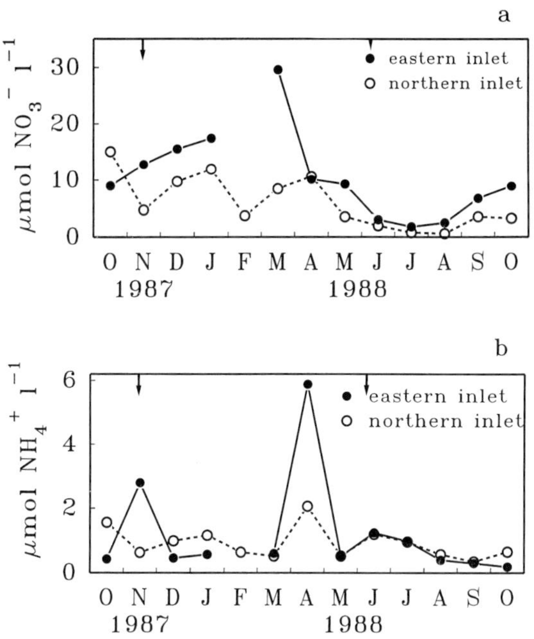

Fig. 8. Annual variability of NH_4^+ and NO_3^- in the two main inlets of Lake Bacivèr. Both inflows drain similar catchment areas but water from one of the streams comes from higher lakes. Arrows indicate the ice cover period.

retention of this ion in the higher lakes. On the other hand, NH_4^+ was similar in the two inlets and much lower than in precipitation, with the exception of a massive release during the snow melt. This increase was lower in the Northern inlet, which suggests there was some consumption of NH_4^+ in the lakes above. Note that the NO_3^- peak occurred one month before the NH_4^+ peak (fig. 8).

In LB, NO_2^- were always very low and the seasonal pattern followed NH_4^+. NO_2^- were higher in LR and showed an appreciable decrease during thaw because of the dilution due to the inflow water from melting snow (0.02 μmol l^{-1}). During summer stratification, the vertical distribution of NO_2^- showed a clear discontinuity in the thermocline (Fig. 9), the concentration in the epilimnion being higher than in the hypolimnion. The difference between the two layers appeared soon and persisted during the whole summer for both years. The cause of this distribution is not easy to ascertain. In oligotrophic pelagic environments, NO_2^- usually increase around the maximum of chlorophyll (Margalef, 1983) or at the base of the photic zone (McCarthy, 1980) However, in our case, the highest value appeared in the layer with least chlorophyll and most light.

The variability of DON in LR points to the importance of the activity of the microorganisms that can use these compounds (bacteria and flagellates) in these cold waters. During melting, there was a dilution due to entering water and, roughly, the DON varies inversely to DIN ($NO_3^- + NH_4^+ + NO_2^-$) (Fig. 7). At the end of the

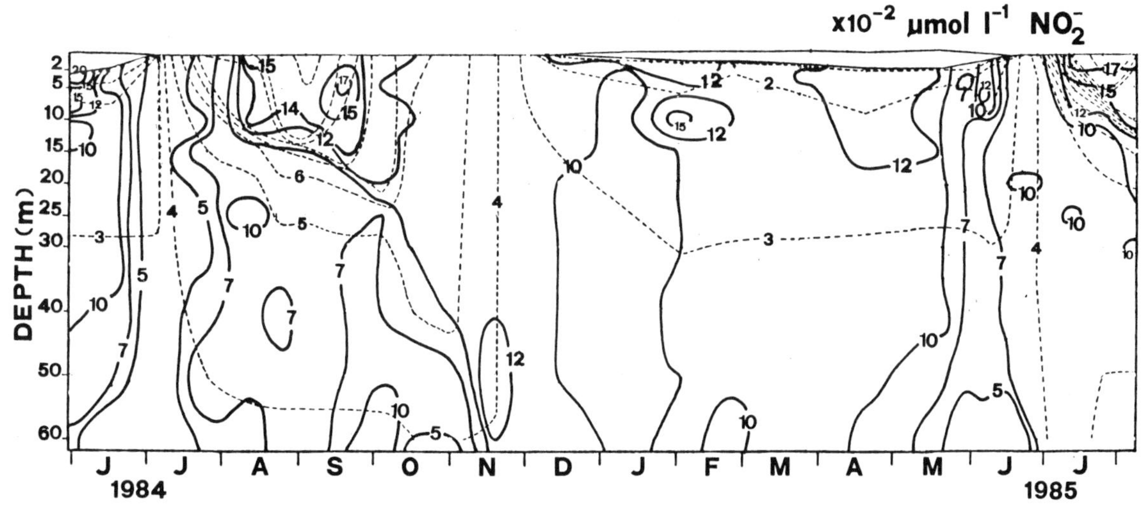

Fig. 9. Isopleth diagram for NO_2^- in Lake Redó. Dashed lines indicate isotherms.

autumn overturn and during the first month of the ice-covered period there was a clear sequence of successive increase and decrease of NO_3^-, DON and NH_4^+. The increase in DON occurred when snow accumulated enough to appreciably decrease the light reaching the water column; during the following month the decrease in DON was parallel to an NH_4^+ increase (Catalan, 1992). Under the ice, the highest values of DON were always associated to zones with more biomass, except during the middle of the winter, when there was a significant increase from 30 m to the bottom.

Nitrogen balance in lake Bacivèr

During most of the year inflow DIN was higher than or equal to outflow DIN (Fig. 10). Only immediately after the thawing of the cover was the outflow more concentrated than the inflow, because of the recirculation of the nitrogen released from the sediment when the flow was interrupted (February and March) and of the nitrogen stored in the ice and snow from atmospheric precipitation. During July and August the values reached a minimum, coinciding with a maximum of PN in the output, which represented an export of a large portion of the planktonic production, because PN inputs at this point were appreciably lower (Fig. 10).

A mass balance can be approximated using mean concentration of the input and output water weighted by the mean monthly discharge of a stream located in the same area. The mean concentration of DIN was 8.22 μmol l^{-1} at the input and 7.38 μmol l^{-1} at the output, thus retention was 0.84 μmol l^{-1}. PN were 1.72 μmol l^{-1} and 2.36 μmol l^{-1}, respectively, then 0.64 μmol l^{-1} PN was exported. The balance between retained DIN and exported PN gave a net retention of 0.20 μmol l^{-1} N, or 69 mmol N m^{-2} y^{-1}. This retention was close to the settled PN, which was 65.4 mmols N m^{-2} y^{-1}. The lake thus acts as a nitrogen trap.

PN was a low fraction of the nitrogen pool in the water column throughout the year (<10%), which was dominated by NO_3^- and DON. However, the amount of nitrogen in macrophytes was much greater than in seston. Annual production of macrophytes required nitrogen sources similar to the retained DIN and settled PN (Table 1) (Ballesteros *et al.*, 1989; Gacia & Ballesteros, 1991). Nonetheless, macrophyte biomass in the

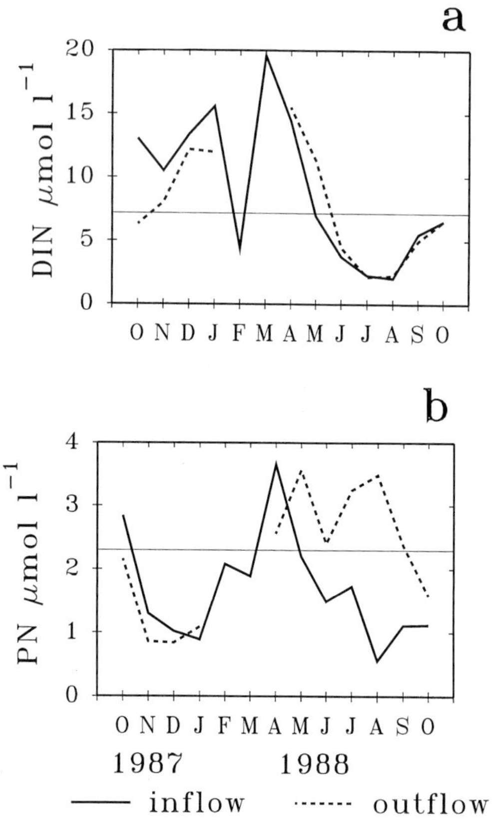

Fig. 10. Annual variability in dissolved inorganic nitrogen and particulate nitrogen in the inflow and outflow of Lake Bacivèr.

Table 1. Nitrogen retained in the biomass of aquatic macrophytes and nitrogen requirements for annual production in Lake Bacivèr.

Community	Biomass mmol N m^{-2}	N Requirement mmol N m^{-2} y^{-1}
Isoetes lacustris <0.8 m	187	75
I. lacustris 0.8–1.8 m	489	84
I. lacustris 1.8–2.3 m	697	70
Sparganium angustifolium	261	260
Nitella gracilis	24	24
Whole lake	326.4	61.7

lake was at steady-state at the annual time scale (Ballesteros *et al.*, 1989). Therefore, there was no net annual requirement. On the other hand, most of the nitrogen for new leaves of *I. lacustris* probably came through internal recycling from old leaves, as it has been described in seagrasses (Borum *et al.*, 1989).

Macrophyte effects on nitrogen compounds in Lake Bacivèr sediments

The main macrophyte species in LB have varied life forms. *Sparganium* is a geophyte with low root:shoot ratio. *Isoetes* is an isoetid (Hutchinson 1975) with high root:shoot (Agami & Waisel 1986), which facilitates the radicular uptake of CO_2, N and P from pore water. Further, *Isoetes* release oxygen through the roots during photosynthetical periods (Sand-Jensen *et al.*, 1982). Thus, different sediment characteristics can be expected for these two communities.

Percentage of nitrogen content in dried sediments was rather high (0.52%–3.23%) compared to North American lakes (0.03%–2.4%, Barko & Smart, 1986) and lakes of similar water chemistry and macrophyte vegetation (0.02%–0.07%, Sand-Jensen & Sondergaard, 1979; 0.014%–1.5%, Chambers & Kalff, 1989). In the surface sediment, nitrogen content was quite similar in the four sampling points, indicating a similar composition of the settled material throughout the lake. Deeper sediment layers showed less nitrogen, especially in the *Isoetes* shallower sampling points (Fig. 11). This decrease might indicate the degree of decomposition of PN undergone at a particular depth at long time scales, assuming no significant changes in the settling matter during this time.

NH_4^+ predominated in the sediments of *Sparganium* beds, and NO_3^- largely dominate the *Isoetes* sediment. NO_3^- content was much lower in the *Sparganium* sediment than in all the *Isoetes* points (Fig. 11). In the latter sediments, NO_3^- presented a maximum in the upper layer, except in the deepest sampling point, where the maximum was at the 3–6 cm deep layer (Fig. 11).

Sparganium sediment showed higher content of

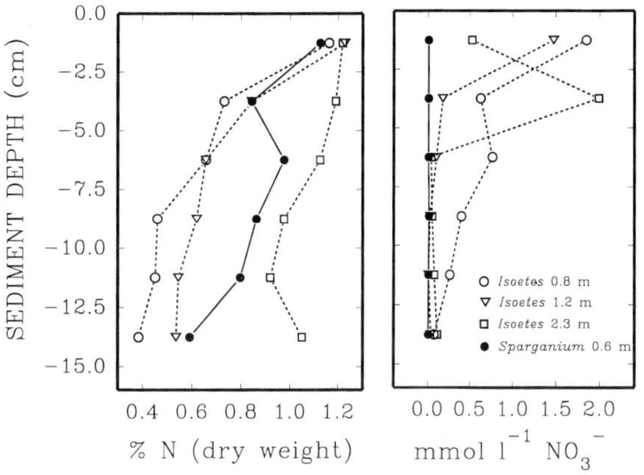

Fig. 11. Percentage of nitrogen in dry sediment and NO_3^- concentration in pore water of macrophyte beds at different depths in Lake Bacivèr.

NH_4^+ (83.5 ± 24.7 μmol l^{-1}) than any of the *Isoetes* profiles (37.6 ± 11.04 μmol l^{-1}). Values in the *Sparganium* sediment were in the lower range for similar environments (120–200 μmol l^{-1}, Nichols & Keeney, 1976; 114–3250 μmol l^{-1}, Barko & Smart, 1986); and NH_4^+ levels in the *Isoetes* sediments were only equivalent to those found in the nitrogen poor terriferrous sediments like in the estuary of Great Bay, New Hampshire, USA (20–40 μmol l^{-1}, Shon, 1987), an order of magnitude lower than the low range for most lakes. However, the NO_3^- levels in the *Isoetes* sediments was equivalent to the NH_4^+ levels of most sediments. This shows the relevance of the oxygen release through the roots of *Isoetes*, which creates and oxidized rhizosphere. The pattern of NO_3^- decrease in the sediment profiles is related to length of the *Isoetes* roots; in the deepest profile the peak moves down to 3–6 cm because of the longer roots of the *Isoetes* plants at this depth (Gacia & Ballesteros, submitted).

There were no significant differences in NO_2^- at increasing depth into the sediments. Mean values were 4.5 ± 0.23 μmol l^{-1} NO_2^- for the *Isoetes* community and 2.03 ± 2.7 μmol l^{-1} for *Sparganium* sediments.

Conclusion

Lakes of the Pyrenees show a broad variability in nitrogen content and in the distribution of the total amount in different oxidation forms. This variability is not directly related to any single or simple combination of physiographical or chemical variables. Only a catchment and lake model based on several physical and biological processes will be able to explain such variability. In our opinion, the following points must be considered in building this model: (1) atmospheric input changes with the distribution of the origin of the storms and the amount of precipitation; (2) NH_4^+ is quickly oxidized or adsorbed in the catchment; (3) aquatic macrophytes significantly modify NO_3^- in the water column (*Ranunculo-Potametum*) or in the pore water of the sediments (*Isoetes*); (4) nitrification depends on pH, and NO_2^- constant values suggest that steady-state is reached during long periods of the year; (5) decomposition in sediments with a similar PN content changes with depth; (6) lakes act as traps for DIN, with uptake of both NH_4^+ and NO_3^- despite the high N:P ratio; (7) DON show a high in-lake variability suggesting fast microbial activities in the water column even in cold water; (8) the melting period is the major event introducing seasonal variability within a lake; differential elution in the snowpack produces an initial peak of nitrogen compounds followed by a dilution of the lake water.

Acknowledgements

This study has been funded by CIRIT, Caixa de Barcelona and CICYT project NAT89-0943.

References

Abrahams, P. W., M. Tranter, T. D. Davies, I. L. Blackwood, P. Brimblecombe & C. E. Vinncent, 1985. The mobilisation of trace elements in a remote Scottish catchment at the onset of snow-melt. Proceedings of the Annual Conference in Trace Substances in Environmental Health, University of Missouri: 67–78.

Agami, M. & Y. Waisel, 1986. The ecophysiology of roots of submerged vascular plants. Physiol. Plant. 24: 607–624.

Ballesteros, E. & E. Gacia, 1991. Una nova associació de plantes aquàtiques als Pirineus: el *Ranunculo eradicati-Potametum alpini*. Bullt. Inst. Cat. Hist. Nat. 59: 82–88.

Ballesteros, E., E. Gacia & L. Camarero, 1989. Composition, distribution and biomass of benthic macrophyte communities from lake Bacivèr, a Spanish alpine lake in the central Pyrenees. Ann. Limnol. 25: 177–184.

Barko, J. W. & R. M. Smart, 1986. Sediment-related mechanisms of growth limitation in submersed macrophytes. Ecology 67: 1328–1340.

Borum, J., L. Murray & M. W. Kemp, 1989. Aspects of nitrogen acquisition and conservation in eelgrass plants. Aquat. Bot. 35: 289–300.

Brimblecombe, P., S. L. Clegg, T. D. Davies, D. Shooter & M. Tranter, 1987. Observations of the preferential loss of major ions from melting snow and laboratory ice. Wat. Res. 21: 1279–1286.

Buishand, T. A., G. T. Kempen, A. J. Frantzen, H. F. Reijnders & A. J. van den Eshof, 1988. Trend and seasonal variation of precipitation chemistry data in the Netherlands. Atmos. Environ. 22: 339–348.

Cadle, S. H., J. M. Dasch & P. A. Mulawa, 1985. Atmospheric concentrations and the deposition velocity to snow of nitric acid, sulphur dioxide, and various particulate species. Atmos. Environ. 19: 1819–1827.

Camarero, L. & J. Catalan, 1993. Chemistry of bulk precipitation in the Central and Eastern Pyrenees (NE Spain). Atmos. Environ. 27A: 83–94.

Catalan, J., 1988. Physical properties of the environment relevant to the pelagic ecosystem of a deep high-mountain lake (Estany Redó, Central Pyrenees). Oecologia aquatica 9: 89–123.

Catalan, J. 1989. The winter cover of a high-mountain Mediterranean lake (Estany Redó, Pyrenees). Wat. Resour. Res. 25: 519–527.

Catalan, J., 1992. Evolution of dissolved and particulate matter during the ice-covered period in a deep high mountain lake. Can. J. Fish aquat. Sci. 49: 945–955.

Catalan, J., E. Ballesteros, E. Gacia, A. Palau & L. Camarero, 1993. Chemical composition of disturbed and undisturbed high-mountain lakes in the Pyrenees: a reference for acidified sites. Wat. Res. 27: 133–141.

Catalan J. & L. Camarero, 1991. Ergoclines and biological processes in high mountain lakes: Similarities between summer stratification and the ice-forming periods in Lake Redó (Pyrenees). Verh. int. Ver. Limnol. 24: 1011–1015.

Chambers, P. & J. Kalff, 1987. Light and nutrients in the control of aquatic plant community structure. 1. *in situ* experiments, J. Ecol., 75: 611–619.

Gacia, E., 1992. Els macrofits submergits dels estanys pirinencs: estructura i dinamica de les poblacions de l'estany Baciver. Thesis Depart. of Ecology, University of Barcelona, 176 pp. [In Catalan].

Gacia, E. & E. Ballesteros, 1991. Two methods to estimate leaf production in *Isoetes lacustris*: A methodological critical assessment. Verh. int. Ver. Limnol. 24.

Gacia, E. & E. Ballesteros, submitted. Population and individual variability of *Isoetes lacustris* L. with depth in a Pyrenean lake.

Grasshoff, K., M. Ehrhardt & K. Kremling, 1983. Methods of

seawater analysis. Verlag Chemie, Weinheim, Germany, 419 pp.

Grennfelt, P. & H. Hultberg, 1986. Effects of nitrogen deposition on the acidification of terrestrial and aquatic ecosystems. Wat. Air Soil Pollut. 30: 945–963.

Hutchinson, G. E., 1975. A treatise on Limnology. III Limnological Botany. Wiley, New York, NY, 660 pp.

Johannessen, M., R. Mosello & H. Barth, 1990. Air Pollution Research Report 20. Commission of the European Communities. Directorate General for Science, Research and Development. Brussels.

Keller, H. M., P. Klöti & F. Forster, 1987. Event studies and the interpretation of water quality in forested basins. Proc. Int. Symposium on acidification and water pathways. Bolkesjo, Norway, May 1987: 237–248.

Levat, Y., P. Lasserre & P. LeCorre, 1990. Seasonal changes in pore water concentrations of nutrients and their diffusive fluxes at the sediment-water interface. J. exp. mar. Biol. Ecol. 35: 135–160.

Margalef, R., 1983. Limnología. Omega, Barcelona.

McCarthy, J. J., 1980. Nitrogen, In I. Morris (ed.), The physiological ecology of phytoplankton. Blackwell, Oxford.

Meesenburg, H., 1987. Acid deposition in the Black Forest. An overview. Documenta Ist. Ital. Idrobiol. 14: 73–82.

Morris, E. M. & A. G. Thomas, 1985. Preferential discharge of pollutants during snowmelt in Scotland. J. Glaciol. 31: 190–193.

Mosello, R., G. A. Tartari & A. Marchetto, 1987. Alterazioni delle deposicioni atmosferiche ed effetti sulle acque superficiali: la situazione dell'Italia Nord Occidentale. Documenta Ist. Ital. Idrobiol. 14: 1–18.

Mosello, R., G. Tartari, B. Sulis & A. Boggero, 1988. Richerche sulle deposicioni acide e sull'acidificacione delle acque superficiali. Acqua-Aria. 1: 61–67.

Mosello, R., G. Tartari & G. A. Tartari, 1985. Chemistry of bulk deposition at Pallanza (northern Italy) during the decade 1975–84. Mem. Ist. ital. Idrobiol. 43: 311–332.

Nichols, D. S. & D. R. Keeney, 1976. Nitrogen nutrition of *Myriophyllum spicatum*: uptake and translocation of N by shoots and roots. Freshwat. Biol. 6: 145–15.

Psenner, R., 1987. Deposizioni acide in Austria ed effetti sulle acque dolci e sulla vegetazione. Documenta Ist. Ital. Idrobiol. 14: 35–71.

Psenner, R., 1989. Chemistry of high mountain lakes in siliceous catchments of the Central Eastern Alps. Aquat. Sci. 51: 1015–1621.

Sand-Jensen, K. & M. Sondergaard, 1979. Distribution and quantitative development of aquatic macrophytes in relation to sediment characteristics in oligotrophic Lake Kalgaard, Denmark. Freshwat. Biol. 9: 1–11.

Sand-Jensen, K., C. Prahl & H. Stokholm, 1982. Oxygen release from roots of submerged aquatic macrophytes. Oikos 39: 349–354.

Schuurkes, J. A. A. R. & R. Mosello, 1988. The role of external ammonium inputs in freshwater acidification. Schweiz. Z. Hydrol. 50: 71–86.

Slanina, J., W. A. Lingerak & L. Bergman, 1976. A fast determination of nitrate in rain and surface waters by means of ultraviolet spectrophotometry. Z. Anal. Chem. 280: 365–368.

Solórzano, L., 1969. Determination of ammonia in natural waters by the phenolhypochlorite method. Limnol. Oceanogr. 14: 799–801.

Young, J. R., E. C. Ellis & G. M. Hidy, 1988. Deposition of air-borne acidifiers in the Western environment. J. Envir. Qual. 17: 1–26.

Hydrobiologia **274**: 29–35, 1994.
J. Fott (ed.), Limnology of Mountain Lakes.

Plankton dynamics in a high mountain lake (Las Yeguas, Sierra Nevada, Spain). Indirect evidence of ciliates as food source for zooplankton

L. Cruz-Pizarro [1], I. Reche [2] & P. Carrillo [2]
[1] *Instituto de Agua, Universidad de Granada, 18071 Granada, Spain;* [2] *Departamento Biología Animal y Ecología, Facultad de Ciencias, Universidad de Granada, 18071, Granada, Spain*

Key words: phytoplankton, zooplankton, protozoan ciliates, seasonal dynamics, high mountain lakes

Abstract

A detailed sampling programme during the ice-free season (July–September) in the oligotrophic lake Las Yeguas (Southern Spain) has shown a well-defined time lag between phytoplankton and zooplankton maximum standing stocks, the former displaying a peak (23 $\mu gC\ l^{-1}$) just after the ice-melting, and the latter by the end of September (80 $\mu gC\ l^{-1}$).

A ratio of autotrophs to heterotrophs lower than 1 which lasted more than two thirds of the study period may suggest a high algal productivity per unit of biomass. The estimated strong top-down regulation of phytoplankton by zooplankton indicates an efficient utilization of resources.

A comparative analysis between the available food supply and the critical food concentration that is necessary to maintain the population of *Daphnia pulicaria* (which constitutes up to 98% of the heterotrophic biomass) proves this species to be food-limited in the lake under study.

To explain the dominance (and development) of such large-bodied cladoceran population, we discuss the possibility of the utilization of naked protozoan ciliates (Oligotrichidae) as a complementary high quality food source, and the exploitation of benthic resources through a coupled daily migration behaviour.

Introduction

Oligotrophy promotes the existence of small sized phytoplankton species with high metabolic rates (Reynolds, 1984; Rott, 1988; Psenner & Zapf, 1990). At the same time it promotes large-bodied zooplankton with low metabolic rates per unit weight (Taylor, 1984). This results, among others, in a simplification of the planktonic community which makes such systems suitable to study complex interactions (Neill, 1988).

The activity of the microbial components in the recycling of nutrients in the euphotic zone of these nutrient-limited ecosystems is likely to be crucial for support of the growth and production of all components in the pelagic community (Axler *et al.*, 1981; Scavia & Laird, 1987; Stockner & Porter, 1988).

In this sense, bacterivory of ciliates that can bypass at least one step in the bacteria to flagellate to ciliate microbial loop (Azam *et al.*, 1983), has been considered very important in providing the feedback of nutrients and making a substantial proportion of the bacterial production available to higher order consumers (Sherr & Sherr, 1987).

In this paper we describe the seasonal variations in the structure of the planktonic community in an oligotrophic high mountain lake and discuss the role of ciliates as a link, moving pico-

and nanoplankton carbon to macrozooplankton, thus allowing the maintenance of a high heterotrophic biomass dominated by large cladocerans.

The possibility of a diel feeding rhythm coupled with the vertical migration of the zooplankton (Lampert & Taylor, 1985) is also considered.

Study site

Las Yeguas is a small (2530 m^2) and shallow (maximun depth: 8 m) oligotrophic lake (chlorophyll concentration range: 0.6–1.76 $\mu g\ l^{-1}$, TP under 30 $\mu g\, l^{-1}$ and high transparency) located at 2800 m in the Sierra Nevada (Southern Spain).

In the early sixties a dam was built to use the lake as a drinking water supply reservoir for a nearby ski station and despite this it failed in increasing the water storage because of some leakage problems, which was presumably responsible for a change in the water quality and phytoplankton composition (Sanchez-Castillo *et al.*, 1989). In fact, Reche (1991) has detected alkalinity values close to 1 meq l^{-1}, far from the range of concentrations measured for the whole Sierra Nevada lakes (between 50 and 400 $\mu eq\ l^{-1}$: Morales-Baquero *et al.* 1992).

Biological communities are rather simple. *D. pulicaria* dominates, in terms of biomass in the zooplankton, whereas the population of *Mixodiaptomus laciniatus* is poorly represented, in contrast with its specific dominant status in more than 70% of the lakes in the Sierra Nevada. Green algae, particularly *Chlorella*, *Oocystis*, *Chlamydomonas* and *Chlorogonium* species all account for about 40–70% of the autotrophic biomass. There are no fish.

Materials and methods

The plankton samples were taken at 2 m intervals from surface to bottom with a double Van Dorn bottle (8 litres each) at a deep central station ($z = 8$ m). Samples were taken twice a week, during the ice-free period of 1986.

A 100 ml subsample from the water bottle was preserved with acid Lugol's solution for the analysis of phytoplankton and protozoans. 50 ml subsamples were sedimented for 48 hours in a 2.6 cm diameter compound chamber, and cells counted in 100 randomly selected fields of view at a magnification of 1000× under an inverted microscope as Sandgren & Robinson (1984) recommend.

For every sample 20 cells of each species were measured to estimate cell volume from the appropriate geometric shape. Biovolume ($\mu m^3\ ml^{-1}$) for each individual taxon was thus determinated by multiplying mean cell volume by cell population density (Vanni, 1987). To get biomass (wet weight) values, a specific gravity of 1 for the phytoplankton, and 1.025 for ciliates (Sorokin & Paveljeva, 1972), was assumed.

The carbon content of the phytoplankton community was estimated using the equations of Strathmann (1967), that from Smetacek (1975) for armoured dinoflagellates, and that proposed by Sorokin & Paveljeva (1972) for the ciliates.

The abundance of zooplankton was obtained by sieving 8 l of water through a 45 μm mesh size net; zooplankton was immediately preserved in 4% formaldehyde. Counting for whole samples was done under an inverted microscope at 150× magnification.

Zooplankton biomass was estimated using length-weight regressions, following Dumont *et al* (1975) and Botrell *et al.* (1976) and then expressed in terms of carbon using the conversion factors proposed by Lampert (1984).

For calculating whether or not food availability was limiting to zooplankton growth, we have followed the method provided by Huntley & Boyd (1984) who suggest that below some critical particulate carbon concentration (C_c:$\mu gC\ l^{-1}$) food would be insufficient to support the predicted maximum growth rate of zooplankton not limited by food availability.

Results

As observed in most of similarly nutrient limited systems, the period just after the ice-melting is characterized by a peak in the phytoplankton biomass (Carrillo, 1989; Carrillo *et al.*, 1990) promoted by both improvements in the physical (increased irradiation) and chemical (readily available nutrients accumulated on the snow during the winter) conditions (Barica & Armstrong, 1971; Larson, 1973).

During the ice-free period the evolution of the algal abundance and biomass depict a well defined pattern, showing a gradual and sustained decrease of such parameters from 11 000 cell ml^{-1} and 23 μgC l^{-1} down to values close to 50 cell ml^{-1} and 0.8 μgC l^{-1} respectively (Fig. 1)

The autotrophic community was dominated at the early stages of the study period by small, non-motile cells belonging to the genus *Chlorella* (2–4 μm) and *Oocystis* (6–8 μm) which contributed between 78 and 90% to the photosynthetic biomass (Fig. 2).

Flagellates were present throughout the entire

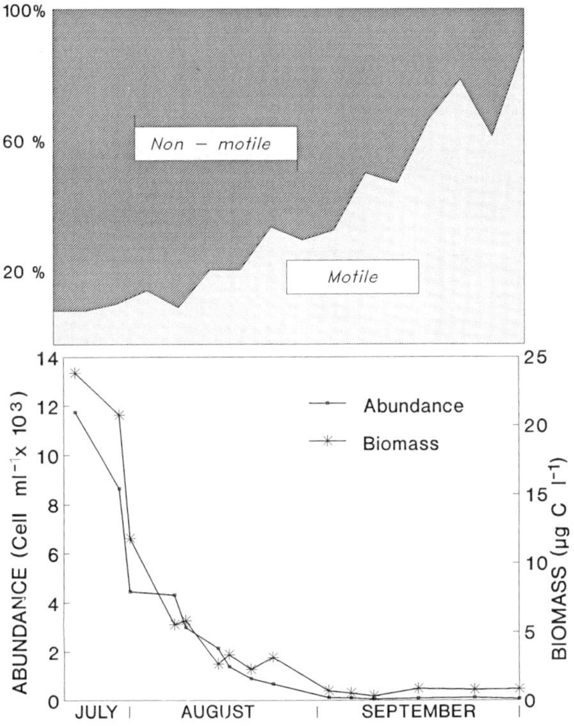

Fig. 1. Evolution of the contribution by motile and non-motile algal fraction to the autotrophic carbon content (top) and of the phytoplankton abundance and biomass (down) in the study period.

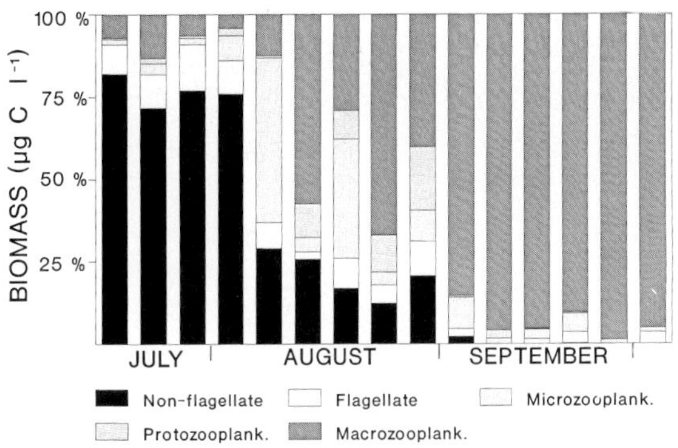

Fig. 2. Relative contribution of the different functional and size groups to the total biomass.

period and showed little variations both in density and in biomass, although at the end of the summer they accounted for more than 90% of carbon available for zooplankton (Fig. 1). In this sense, Fig. 2 shows a more realistic image, since it demonstrates that flagellates did not contribute significantly to the plankton biomass.

The development of the zooplankton biomass basically performed a reversed pattern to that described by algae, rising to the maximum values (80 μgC l^{-1}) by the end of September (Fig. 3a).

A detailed analysis of the heterotrophic community allows to establish a clear seasonal succession for the different groups involved. In the first period, and for about two to three weeks, rotifer populations (*Hexarthra bulgarica* and *Euchlanis dilatata*) did develop, reaching biomass values up to 8 μgC l^{-1}, just coinciding with the greatest decrease of the algal populations (Figs 3b and 1).

A second stage was dominated by the protozooplankton, mainly small (10–25 μm) naked oligotrichidae ciliates, at cellular densities (2–10 cell ml^{-1}) within the range usually found in similar oligotrophic environments (Gates & Lewg, 1984; Carrick & Fahnenstiel, 1990), which in terms of carbon (up to 7 μgC l^{-1}) represent an important contribution to the actual biomass.

Finally, during September, the heterotrophic community was largely dominated by the macrozooplankton, increasing in biomass throughout the ice-free period from 1 μgC l^{-1} in July to 79 μgC l^{-1} at the end of September, thus resem-

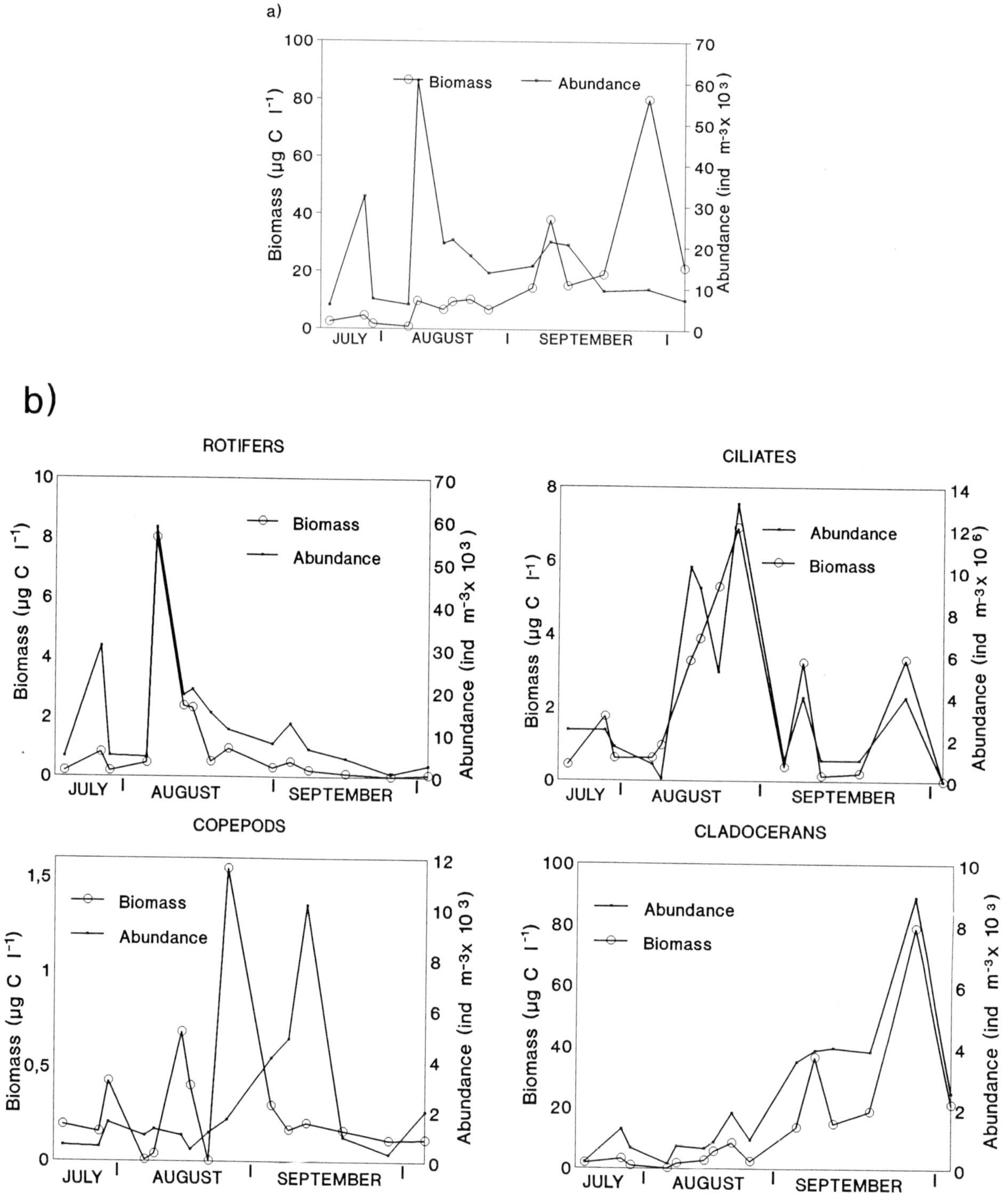

Fig. 3. Changes in the total zooplankton biomass and abundance (microzooplankton + macrozooplankton) (a), and of the different taxonomic groups (b) during the ice-free period.

bling the pattern described for the whole zooplanktonic assemblage.

This fraction, comprised in more than 90% by great sized cladocerans (*D. pulicaria*) accounted, during the study period, for between 5 and 90% of the particulate pelagic carbon (Fig. 2) and between 18 and almost 98% if only the zooplankton community is taken into account.

The only two copepod species in the lake (*Mixodiaptomus laciniatus, Eucyclops serrulatus*) are poorly represented. In fact, the former one, which dominates the zooplankton biomass during most of the ice-free period in other Sierra Nevada lakes (Carrillo, 1989) never surpassed 10% of the heterotrophic biomass in lake Las Yeguas.

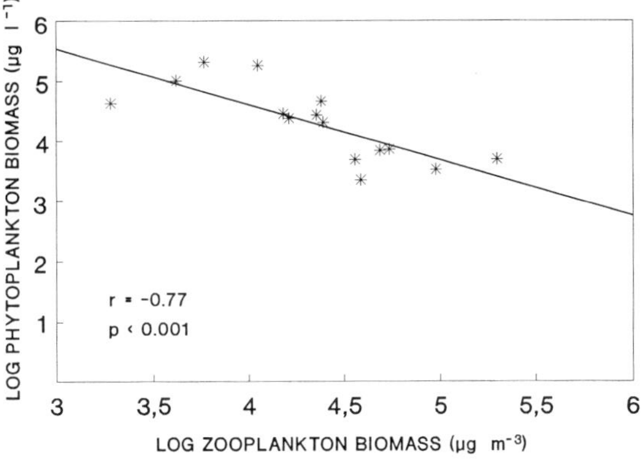

Fig. 4. Regression of phytoplankton biomass on zooplankton biomass (log-log).

Discussion

The analysis of the changes in biomass and in the trophic structure of the pelagic community during the investigation period has shown that for more than two-thirds of this time the autotrophs: heterotrophs ratio was lower than 1 (see for instance Figs 1 and 3a). This is indicative of a phytoplankton assemblage with relatively low standing stock and high turnover rates (Stegmann & Peinert, 1984), but to explain the rather high increase in the carbon content (three and fourfold) between both trophic levels, a high zooplankton efficiency should be addressed.

In fact, the zooplankton biomass (independent var.) *vs* phytoplankton biomass (dependent var.) regression analysis, shows a reverse and highly significant relationships between both variables (Fig. 4), suggesting a control of the algal biomass by the zooplankton, i.e. a top-down mechanism seems to be the main factor responsible for the primary producers regulation. These results agree with those predicted for oligotrophic systems (McQueen *et al.*, 1986) and for environments dominated by large cladocerans (McQueen *et al.*, 1989).

Even accepting such high phytoplankton turnover and zooplankton efficiency rates it is still difficult to approve that the development of *Daphnia pulicaria*, a species with high food requirements (Kasprzak *et al.* 1986), may only be maintained by such scarce resources.

In Fig. 5 we have compared the calculated critical food concentration required for this species (while considering an exclusive herbivorous behaviour) and the food amount available from the autotrophic fraction. According to our data, *D. pulicaria* should be food limited except for the time immediately after thaw.

Ciliated protozoans have been considered as an important link in aquatic systems by feeding on size particles not efficiently grazed by large zooplankton and serving, in turn, as readily as-

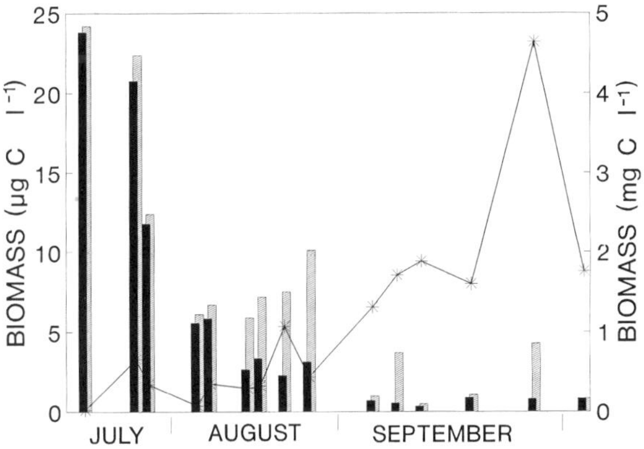

Fig. 5. Critical food concentration estimated for Daphnia pulicaria (mg C l^{-1}, solid line) and available food quantity (µg C l^{-1}, black bars: algae, striped bars: algae + ciliates) during the study period.

similated prey for such grazers (Porter *et al.*, 1979; Beaver & Crisman, 1982).

In lake Las Yeguas the protozoan contribution to the available carbon source for consumers was highest in the second half of August. Even then it did not exceed 70% (between 3 and 7 μgC l^{-1}), very far from the estimated critical food concentration for *D. pulicaria* (between 0.43 and 1.1 mgC l^{-1}) in the same period (Fig. 5).

The importance of protozoans as prey, however, may be seen in their high nutritional value: high energy and protein content (Stoecker & Capuzzo, 1990; DeBiase *et al.*, 1990; Gifford, 1991).

D. pulicaria is probably the only large macrozooplankter in the lake that can effectively utilize all components of the microbial communities ranging in size from 1–50 μm (Stockner & Porter, 1988), thus being able to outcompete calanoids which, in such oligotrophic lakes, cannot effectively graze on picoplankton or very small nanoplankton particles (Scavia & Laird, 1987). In this respect, the lack of ability of cladocerans species to use taste to discriminate food quality should also be considered (DeMott,1986).

In contrast, the high transparency of the water and the shallowness of the lake allow the development of an important epipelic community consisting of Diatoms, Euglenophyceae together with some Zygnematales species which have represented up to 4.6 mgC l^{-1} in samples taken close to the maximum depth of the lake, after disturbing the bottom (Reche, in prep.). This extremely high food reservoir might be available for *Daphnia* as Echevarria *et al.*(1990) suggest, through a coupled diurnal feeding rhythm and a vertical migration, widely described for this species in a rather similar ecosystem (Cruz-Pizarro, 1978; 1981; Carrillo *et al.*, 1991).

Acknowledgements

This research was supported by CICYT Project Nat 91-570 and a FPI grant to I. Reche. We thank Dr Sanchez-Castillo his advice with the algal taxonomy and Dr V. Korínek for his determination of the taxonomic status of the *Daphnia* population. The authors are indebted to Dr J. Fott for his invaluable comments in the review of the manuscript.

References

Axler, R. P., G. W. Redfield & C. R. Goldman, 1981. The importance of regenerated nitrogen to phytoplankton productivity in a subalpine lake. Ecology 62: 345–354.

Azam, F., T. Fenchel, J. G. Field, J. S. Gray, L. A. Meyer-Reil & F. Thingstad, 1983. The ecological role of water-column microbes in the sea. Mar. Ecol. Prog. Ser. 10: 257–263.

Barica, J. & F. A. Armstrong, 1971. Contribution by snow to the nutrient budget of the small Northwestern Ontario Lakes. Limnol. Oceanogr. 16: 891–899.

Beaver, J. R. & T. L. Crisman, 1982. The trophic response of ciliated protozoans in freshwater lakes. Limnol. Oceanogr., 27: 246–253.

Botrell, H. H., A. Duncan, Z. M. Gliwicz, E. Grygierek, A. Herzig, A. Hillbricht-Ilkowska, H. Kurasawa, P. Larsson & T. Weglenska, 1976. A review of some problems in zooplankton production studies. Norw. J. Zool. 24: 419–456.

Carrick, H. J. & G. L. Fahnenstiel, 1990. Planktonic protozoa in lakes Huron and Michigan: seasonal abundance and composition of ciliates and dinoflagellates. J. Great. Lakes Res. 16: 319–329.

Carrillo, P., 1989. Analisis de las interacciones tróficas en un sistema oligotrófico. Ph D. Thesis Univ. Granada, 212 pp.

Carrillo, P., L. Cruz-Pizarro & R. Morales-Baquero, 1990. Effects of unpredictable atmospheric allochtonous input on the light climate of an oligotrophic lake. Verh. int. Ver. Limnol. 24: 97–101.

Carrillo, P., P. Sanchez-Castillo & L. Cruz-Pizarro, 1991. Coincident zooplankton and phytoplankton diel migration in a high mountain lakes (La Caldera, Sierra Nevada, Spain). Arch. Hydrobiol. 122: 57–67.

Cruz-Pizarro, L., 1978. Comparative vertical zonation and diurnal migration among Crustacea and Rotifera in the small high mountain lake La Caldera (Granada, Spain). Verh. int. Ver. Limnol. 20: 1026–1032.

Cruz-Pizarro, L. 1981. Estudio de la comunidad zooplanctónica de un lago de alta montana (La Caldera, Sierra Nevada, Granada). Ph. D. Thesis. Universidad de Granada, 186 pp.

DeBiase, A. E., R. W. Sanders & K. G. Porter, 1990. Relative nutritional value of ciliate protozoa and algae as food for Daphnia. Microb. Ecol. 19: 199–210.

DeMott, W. R., 1986. The role of taste in food selection by freshwater zooplankton. Oecologia (Berlin), 69: 334–340.

Dumont, H. J., I. Van de Velde & S. Dumont, 1975. The dry weight estimate of biomass in a selection of cladocera, copepoda and rotifera from the plankton, periphyton and benthos of continental waters. Oecologia 19: 75–97.

Echevarria, F., P. Carrillo, F. Jimenez, P. Sánchez-Castillo, L. Cruz-Pizarro & J. Rodriguez, 1990. The size-abundance distribution and taxonomic composition of plankton in an oligotrophic, high mountain lake (La Caldera, Sierra Nevada, Spain). J. Plankton Res. 12: 415–422.

Gates, M. A. & U. T. Lewg, 1984. Contribution of ciliated protozoa to the planktonic biomass in a series of Ontario Lakes: Quantitative estimates and dynamical relationships. J. Plankton Res. 6: 433–456.

Gifford, D. J. 1991. The protozoan-metazoan trophic link in pelagic ecosystems. J. Protozool. 38: 81–86.

Huntley, M. E. & C. M. Boyd, 1984. Food-limited growth of marine zooplankton. Am. Nat. 124: 455–478.

Kasprzak, P., V. Vyhnálek & M. Straskraba, 1986. Feeding and food selection in Daphnia pulicaria (Crustacea:Cladocera). Limnologica (Berlin) 17: 309–323.

Lampert, W., 1984. The measurement of respiration. In: Downing, J. A. & F. H. Rigler (eds), A manual on methods for the assessment of secondary productivity in fresh water. IBP 17. Blackwell, Oxford: 413–468.

Lampert, W. & B. E. Taylor 1985. Zooplankton grazing in a eutrophic lake: implications of vertical migration. Ecology 66: 68–82.

Larson, G. L., 1973. A limnological study of a high mountain lake in Mount Rainer National Park, Washington State, USA. Arch. Hydrobiol. 72: 10–84.

McQueen, D. J., J. R. Post & E. L. Mills, 1986. Trophic relationships in freshwater pelagic ecosystem. Can. J. Fish. aquat. Sci. 43: 1571–1581.

McQueen, D. J., M. R. S. Johannes, T. J. Stewart & D. R. S. Lean, 1989. Bottom-up and top-down impacts on freshwater pelagic community structure. Ecol. Monogr. 59: 289–309.

Morales–Baquero, R., P. Carrillo, L. Cruz-Pizarro & P. SánchezCastillo. 1992. Southernmost high mountain lakes in Europe (Sierra Nevada) as reference sites for pollution and climate change monitoring. Limnetica 8: 39–47.

Neill, W. E., 1988. Complex interactions in oligotrophic lake food webs: responses to nutrient enrichment. In S. R. Carpenter (ed.), Complex interactions in lake communities. Springer-Verlag, New York.

Porter, K. G., M. L. Pace & J. F. Battey, 1979. Ciliate protozoans as links in freshwater planktonic food chains. Nature 277: 563–565.

Psenner, R. & F. Zapf, 1990. High mountain lakes in the Alps: peculiarities and biology. In Johannessen, M., Mosello, R. & H. Barth (eds), Acidification processes in remote mountain lakes. Air pollution research report 20, Commission of the European Communities, Brussels: 22–37.

Reche, I., 1991. Analisis de la sucesión fitoplanctónica en una laguna de alta montaña: Las Yeguas (Sierra, Nevada). Tesis de Licenciatura. Univ. Granada, 120 pp.

Reynolds, C., 1984. The ecology of freshwater phytoplankton. Cambridge University Press, Cambridge, 384 pp.

Rott, E., 1988. Some aspects of the seasonal distribution of flagellates in mountain lakes. Hydrobiologia 161/Dev. Hydrobiol. 45: 159–170.

Sánchez-Castillo, P., L. Cruz-Pizarro & P. Carrillo, 1989. Caracterizacián del fitoplancton de las lagunas de alta montaña de Sierra Nevada (Granada, Espana) en relación con las caracteristicas fisico-químicas del medio. Limnetica 5: 37–50.

Sandgren, C. D. & J. V. Robinson, 1984. A stratified sampling approach to compensating for non-random sedimentation of phytoplankton cells in inverted microscope settling chambers. Br. Phycol. J. 19: 67–72.

Scavia, D & G. A. Laird, 1987. Bacterioplankton in lake Michigan: dynamics, controls, and significance to carbon flux. Limnol. Oceanogr. 32: 1017–1032.

Sherr, E. B. & B. F. Sherr, 1987. High rates of consumption of bacteria by pelagic ciliates. Nature 325: 710–711.

Smetacek, V., 1975. Die Sukzession des phytoplanktons in der westlichen Kieler Bucht. Ph. D. Thesis, Univ. Kiel, 151 pp.

Sorokin, Yu I. & E. B. Paveljeva, 1972. On the quantitative characteristics of the pelagic ecosystem of Dalnee Lake (Kamchatka). Hydrobiologia 40: 519–552.

Stegmann, P. & R. Peinert, 1984. Interrelations between herbivorous zooplankton and phytoplankton and their effect on production and sedimentation of organic matter in Kiel Bight. Limnologica 15: 487–495.

Stockner, J. G. & K. G. Porter, 1988. Microbial food webs in freshwater planktonic ecosystems. In S. R. Carpenter (ed.), Complex interactions in lake communities. Springer-Verlag, New York: 69–83.

Stoecker, D. K. & J. M. Capuzzo, 1990. Predation on protozoa: its importance to zooplankton. J. Plankton Res. 12: 891–908.

Strathmann, R. R., 1967. Estimating the organic carbon content of phytoplankton from cell volume. Limnol. Oceanogr. 12: 411–418.

Taylor, W. D., 1984. Phosphorus flux through epilimnetic zooplankton from lake Ontario: relationships with body size and significance to phytoplankton. Can. J. Fish aquat. Sci. 41: 1702–1712.

Vanni, J. M., 1987. Effects of nutrients and zooplankton size on the structure of a phytoplankton community. Ecology 68: 624–635.

Hydrobiologia **274**: 37–47, 1994.
J. Fott (ed.), Limnology of Mountain Lakes.

Acidification of lakes in Šumava (Bohemia) and in the High Tatra Mountains (Slovakia)

Jan Fott, Miroslava Pražáková, Evžen Stuchlík & Zuzana Stuchlíková
Department of Hydrobiology, Faculty of Science, Charles University, Viničná 7, 128 44 Prague 2, Czech Republic

Key words: mountain lakes, acidification, phytoplankton, zooplankton

Abstract

Acidification of lakes takes place when pH of rainwater is less than 4.5 and the catchments lie on sensitive geology. Both conditions are met for most lakes in Bohemia and Slovakia. Since 1978 we have studied mountain lakes in the Šumava and in the High Tatra Mountains.

In Šumava the three lakes under study are of glacial origin. The catchments are small, with steep sides covered by spruce. The bedrocks are biotite-rich paragneiss, together with gneiss, quartzite and granite. In summer 1936 surface pH was 5.7–6.9 in the Lake Čertovo and 6.9–7.0 in the Lake Černé. Now the pH values are 4.3–4.8 in the two lakes and in the Lake Prášilské as well. Old reports on zooplankton are from the years 1871, 1892–96, 1935–37, 1947 and 1960. Since 1979 we have not found any planktonic Crustacea in the lakes Černé and Čertovo. Lake Prášilské is inhabited by *Daphnia longispina* and *Cyclops abyssorum*. In July 1989 the pH values were 4.4, 4.7 and 4.7, concentrations of labile monomeric Al were 0.83, 0.68 and 0.24 mg l^{-1} in the lakes Čertovo, Černé and Prášilské, respectively. High levels of toxic Al compounds might be responsible for the extinction of planktonic Crustacea in the lakes Čertovo and Černé. All the three lakes are void of fish at present.

In the High Tatra Mts. we examined more than 40 lakes above timberline in altitudes 1612–2145 m. They are all clearwater, naturally fishless lakes. The bedrock is granite. Owing to different levels of calcium the lakes are now in different stages of acidification. According to recent changes in the zooplankton they can be divided into three groups:

(1) Species composition of planktonic Crustacea has not changed.
(2) Planktonic Crustacea were present until 1973 but are absent now.
(3) From the original species of Crustacea only *Chydorus sphaericus* is present.

The three groups are well separated along the gradients of calcium, ANC and pH. They can be identified with the Henriksen's bicarbonate (our group 1), intermediate (our group 2) and acid (our group 3) lakes. We suppose that in the process of acidification the lakes of the group (2) have been shifted from oligotrophy to ultraoligotrophy.

Introduction

In the Czech and Slovak Republics there are only two regions where lakes of natural origin occur. One of them is the Šumava (Böhmerwald, Bohemian Forest) with 5 small mountain lakes. They are situated in the forested zone at altitudes of about 1000 m. In the second, the High Tatra

Table 1. Morphometry and geology of the three investigated lakes in Šumava. The names of the lakes are in Czech and in German. A′, drainage area; A, lake area; z_{max}, maximum depth, z_m, mean depth.

Lake	A′ (ha)	A (ha)	A/A′	z_{max} (m)	z_m (m)	Geology	Altitude (m)
Černé (Schwarzer See)	129	18.4	0.14	39.8	15.6	Biotite-rich paragneiss	1008
Čertovo (Teufelssee)	88	10.3	0.12	36.7	17.0	Biotite-rich paragneiss and quartzite	1030
Prášilské (Stubenbacher See)	54	3.7	0.07	14.9	7.4	Gneiss granite	1079

Mountains (Slovakia), there are about 130 small lakes at altitudes of 1300–2200 m, most of them above timberline in the alpine zone.

As it is agreed (Henriksen, 1980; Wright *et al.*, 1980), the acidification of lakes takes place whenever the pH of rainwater is less than 4.5 and the catchments lie on sensitive geology. Both conditions are met for the lakes in Šumava and in the Tatras. The annual weighted means of rainwater pH were 4.2–4.5 in the Tatra area in 1977–1981 (Cerovský, 1983). Moldan (1991) gives the annual weighted mean 4.36 as the background value in Bohemia (sampling station Hrádek, 1979–1984). In Šumava, the bedrocks are biotite-rich paragneiss together with gneiss, quartzite and granite. The sensitive part of the Tatras is formed by granite.

The studies on acidification of the lakes in Šumava and in the Tatra Mountains were started in 1978 by a team of the Department of Hydrobiology, Charles University, Prague. Results appeared so far as unpublished reports in Czech or as papers of limited circulation (Stuchlík *et al.*, 1985; Fott *et al.*, 1987; Fott *et al.*, 1992).

Methods

Acid Neutralizing Capacity (Gran titration), conductivity (20 °C) and pH were measured in the laboratory within 24 h after sampling. The major ions were determined by atomic absorption spectrometry and/or by ion-selective electrodes. The major anions were determined according to Mackereth *et al.* (1978), nitrate according to Strickland & Parsons (1968). Calcium, magnesium and sulphate were corrected for sea salts according to Henriksen (1980, Table 2 l.c. world

Table 2. pH in the three lakes in Šumava, May–September. Sources: 1936 – Jírovec & Jírovcová (1937); 1947 – Weiser (1947); 1956–61 – Landa *et al.* (1984) and Procházková & Hrbáček (unpublished); 1979–89 – numerous measurements by the present authors and J. Veselý.

Period	Lake Černé		Lake Čertovo	Lake Prášilské
	Surface	Hypolimnion	Surface	Surface
1936	6.9–7.0	5.7–6.7	5.7–6.9	
Inflow	6.2			
1947	6.2			
1956–61	5.4–6.2	5.6–6.1	4.5–5.0	4.1(?), 5.3–5.8
1979–89	4.4–4.8	4.6–5.0	4.3–4.4	4.3–4.7
Inflows	3.8–4.7		3.8–4.3	4.2–4.7

average). Reactive Al was measured by the method of Dougan & Wilson (1974). The fractionation between labile and non-labile monomeric Al was done as recommended by Røgeberg & Henriksen (1985), but manually with use of a peristaltic pump only. Chlorophyll *a* was measured according to Lorenzen (1967) or fluorimetrically after calibration by the previous method. Organic carbon was measured as reduction capacity, according to Mackereth *et al.* (1978). Particles were retained on Whatman GF/C glass fibre filters. Samples of phytoplankton and zooplankton were taken from a boat using a van Dorn or a Schindler sampler, respectively.

Results and discussion

Lakes in Šumava

Basins of the lakes in the region Šumava–Bayerischer Wald were excavated by single small glaciers during the last glaciation. Their common features are a cirque with a steep 'lake wall' falling from the ridge to the lake, and a lake moraine. The morphometrical data on the three lakes treated in this study are given in Table 1. The relatively small catchments with steep sides are covered by coniferous forest composed mostly of Norwegian spruce. Rational forestry management started in the mid-19th century, replacing a period of uncontrolled exploitation of the primeval forest. Since then no major change in the vegetation cover has taken place.

Old pH data in the three lakes, together with recent data, are summarized in Table 2. Although the old pH data based on indicator measurements are not so reliable as the recent, the drop in pH from the first measurements in 1936 until now is beyond dispute. Care must be taken, however, in the case of the Lake Černé, which was influenced to an unknown degree by the performance of a small hydroelectric power station built in 1930. The station was pumping water from the river Úhlava back up to the lake. The mean residence time was shortened from about three years to some half a year and the onset of the recent acidification could have been retarded until 1975, when the pumping was discontinued. No such activities were performed on the other lakes. The Lake Čertovo was more acid than the Lake Černé, but it is not clear whether all the difference was brought about by the pumping activity of the power station in 1930–1975 or not. The pH differences among the three lakes are small at present, although the Lake Čertovo is still the most acid. Arzet (1987) investigated the diatom flora in the sediment of the Lake Čertovo, and using the methods of pH inference from diatom taxa he concluded that the praeacidification pH was 4.8 to 5.0. These values are by 1–2 units lower than those actually measured in 1936; underestimation of pH inferred from fossil diatoms was mentioned by Schindler (1988).

The chemistry of acidification of the lakes Černé and Čertovo was described recently by Veselý & Majer (1992).

Phytoplankton of the three lakes are similar and poor in number of species, most of the species are flagellates (Table 3). Vertical profiles of chlorophyll and densities of dominant species are

Table 3. Algae frequently found in phytoplankton of the lakes Černé, Čertovo and Prášilské (the dominant species marked *). Except *Bitrichia* and *Raphidonema* all the species are flagellates.

Chrysophyceae	* *Bitrichia ollula* (Fott) Bourelly
	* *Chromulina* sp.
	* *Dinobryon pediforme* (Lemmermann) Steinecke
	Mallomonas sp.
	Synura echinulata Korschikov
Cryptophyceae	*Cryptomonas erosa* Ehrenberg
	Cryptomonas gracilis Skuja
	Cryptomonas marssonii Skuja
Dinophyceae	* *Gymnodinium uberrimum* (Allman) Kofoid et Swezy
	Katodinium bohemicum (Fott) Litvinenko
	* *Peridinium umbonatum* Stein (= *P. inconspicuum* Lemmermann)
Chlorophyceae	* *Carteria* (incl. *Provasoliella*) sp. div.
	Chlamydomonas sp.
	Chlorogonium sp.
	Raphidonema nivalis Lagerheim

Fig. 1a

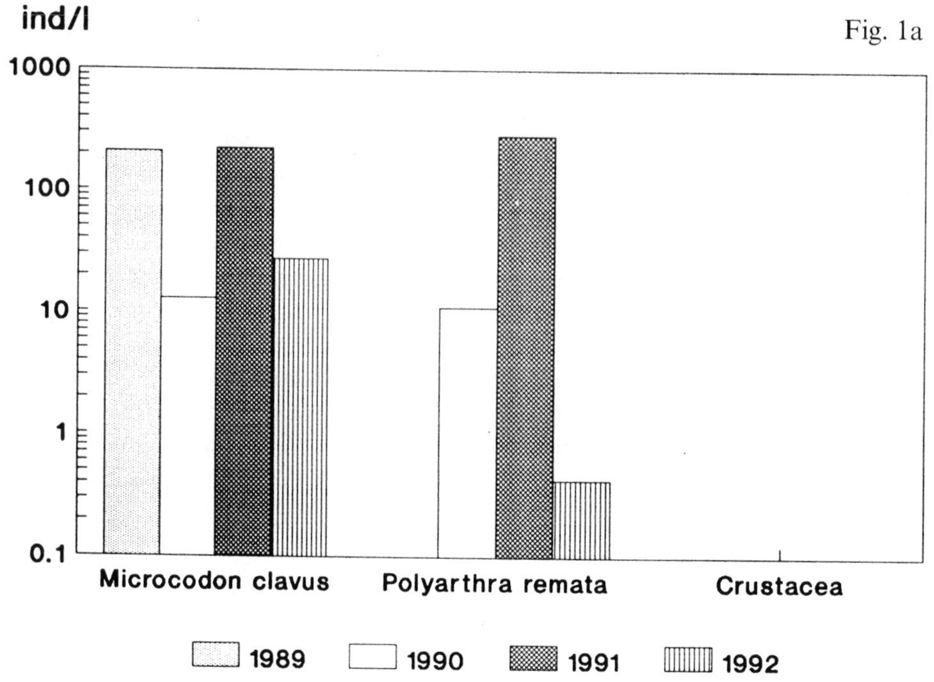

Fig. 1b

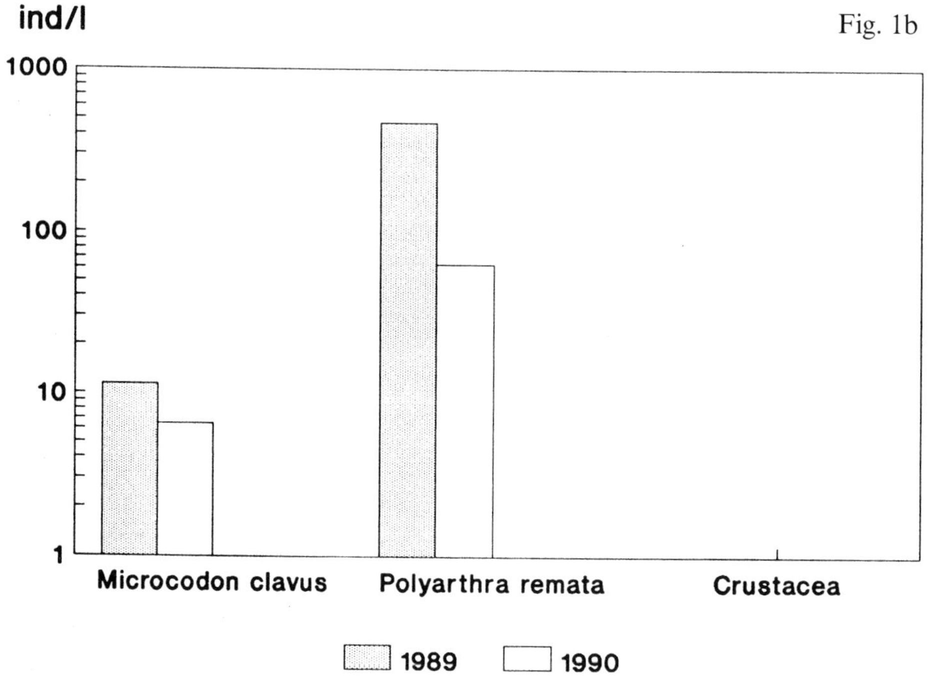

irregular, with peaks at the surface, in the intermediate depth or at the bottom. Summer chlorophyll concentrations are 1–5 $\mu g\ l^{-1}$ in the lakes Černé and Čertovo, and 1–10 $\mu g\ l^{-1}$ in the Lake Prášilské. Surface patches of flagellates (*Carteria* sp. or *Peridinium umbonatum*), making up 40–60 $\mu g\ l^{-1}$ chlorophyll, were observed in the Lake Černé under calm weather in summer.

Fig. 1c

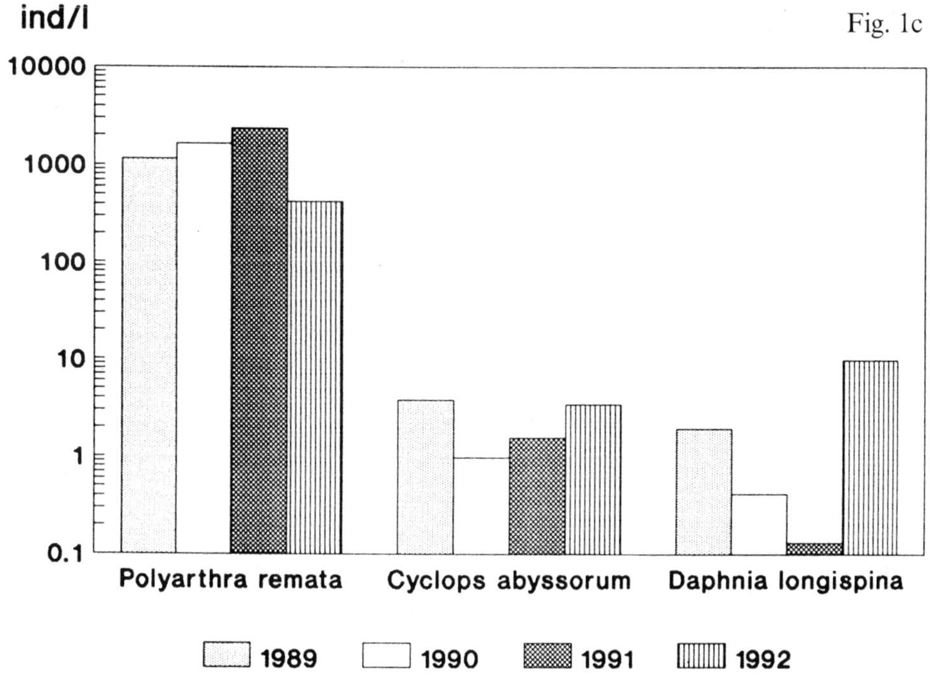

Fig. 1. Summer densities of zooplankton in the Lake Černé (1a), Čertovo (1b) and Prášilské (1c). Volume-weighted means of samples taken on the vertical in mid-summer.

The species composition is similar to that of acidified lakes elsewhere (see Lydén & Grahn, 1985 for a review), but many species were already present in 1935–1936, when B. Fott (unpublished protocols) studied phytoplankton of the Lake Černé. He found *Peridinium umbonatum* and *Dinobryon pediforme* as common species and described *Bitrichia ollula* and *Katodinium bohemicum* as new species (Fott B., 1937, 1938). A species found by him but absent now is *Cyclotella* sp. The comparison of the present state with the old findings proves that many phytoplankton species occurring in acid-sensitive oligotrophic lakes are able to survive when the lake becomes acidic. The

Table 4. Planktonic Crustacea in the Černé Lake, from 1871 to our times. Sources: (1) Frič, 1871. (2) Frič & Vávra, 1897. (3) Šrámek-Hušek, 1942. (4) Weiser, 1947. (5) Hrbáček, unpublished. (6) our data.

Source Period	(1) 1871 ~	(2) 1892– 1896	(3) 1935– 1937	(4) 1947 ~	(5) 1960 ~	(6) 1979– 1989
Holopedium gibberum	+	+		+		
Daphnia longispina	+	+	*			
Ceriodaphnia quadrangula		+	+	+	+	**
Bosmina longispina	+	+				
Acanthodiaptomus denticornis	+					
Cyclops abyssorum	+	+	+	+	+	

* 2 specimens found in 1935
** 1 specimen found in 1979
~ one sample.

high abundance of *Peridinium umbonatum* in lakes with no crustacean zooplankton and low densities of rotifers (L. Černé & Čertovo) is worthy of attention. It indicates that selective herbivory does not play an important role in common occurrence of *Peridinium umbonatum* (= *P. inconspicuum*, see Popovský & Pfiester, 1990) in acidified lakes. Yan & Strus (1980) and Havens & DeCosta (1985) came to the same conclusion.

Zooplankton research of the lakes began in 1871 when Antonín Frič visited 6 lakes (Černé, Čertovo, Prášilské, Laka, Gr. Arbersee, Kl. Arbersee) and reported lists of cladocerans and copepods (Frič, 1872). Further research was directed mostly to the Lake Černé (Table 4). In 1871 Frič found 5 crustacean species in the plankton and described high biomass of the catch. Such biomass was never found in 1892–1896 when he returned to the lake to carry out thorough studies. Moreover, large *Daphnia* and *Holopedium* were replaced by small *Bosmina longispina* and *Cyclops abyssorum*. Frič & Vávra (1897) explained this by predation of brook trout (*Salvelinus fontinalis*) that was introduced repeatedly to the lake in the 1890's. This is one of the first attempts to explain changes in plankton by fish predation. Šrámek-Hušek (1942) found the zooplankton still more impoverished. No planktonic Crustacea are present in the Lake Černé now.

It could be speculated how changes in the vegetation cover during the last 200 years contributed to this development in the lake. The last apparent change, however, was the introduction of spruce monoculture in the mid-19th century. The ultimate stage of the succession must have been caused by the acid precipitation.

Now, the only zooplankton species in the lakes Černé and Čertovo are the rotifers *Microcodon clavus* and *Polyarthra remata* (Fig. 1a, b). *Microcodon clavus* is a pH-tolerant species (Berzins & Pejler 1987) living normally in *Sphagnum* moss or among water plants (Koste, 1978). In the lakes under study it forms planktonic populations. A similar case of a *Sphagnum* – dwelling rotifer, *Polyarthra minor*, that invaded planktonic zone of acidified lakes, was reported by Morling & Pejler (1990). *Polyarthra remata* is a pH-tolerant species (Berzins & Pejler, 1987) known from acidified lakes in Scandinavia (Almer *et al.*, 1978; Hörnström *et al.*, 1984). From the three lakes only the Lake Prášilské has the planktonic zone inhabited by Crustacea: *Cyclops abyssorum* and *Daphnia longispina* (Fig. 1c). The presence of *Daphnia longispina* in an acid-stressed clearwater lake is contradicting to the experience from Scandinavia (Brett, 1989).

In a study on effects of pH and Al on lake plankton, Hörnström *et al.* (1984) concluded that 'high supply of aluminium to the acid lakes is, through oligotrophication and toxicity, alone responsible for the sparse zooplankton fauna.' We tested the hypothesis that the Lake Prášilské, with *Cyclops* and *Daphnia*, has less labile monomeric Al than the lakes Černé and Čertovo. This is actually the case (Fig. 2). In the summers of 1988

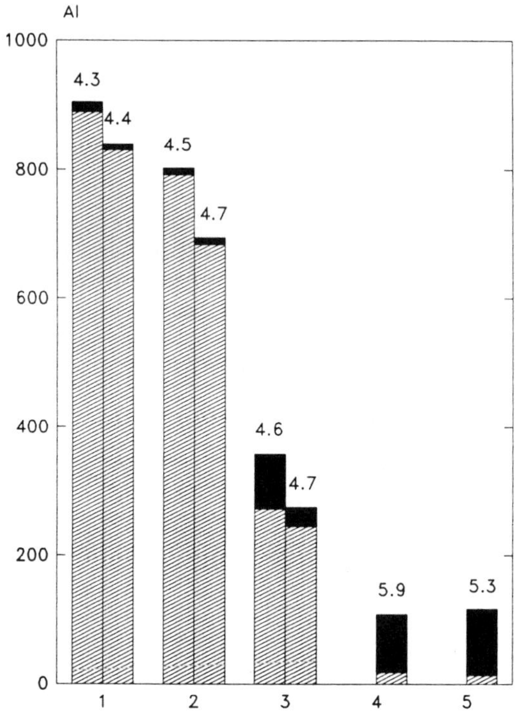

Fig. 2. Labile (striped columns) and non-labile (black columns) aluminium ($\mu g\ l^{-1}$) in mid-summer surface samples from the lakes Čertovo (1), Černé (2), Prášilské (3), Laka (4) and from a brownwater brook (5) near Kvilda. 1, 2, 3: 1988 (1st column), 1989 (2nd column). 4, 5: 1989. Values of pH are indicated on the top.

and 1989 the concentrations were 0.89 and 0.83 mg l^{-1} in the Lake Čertovo, 0.79 and 0.68 mg l^{-1} in the Lake Černé, 0.27 and 0.25 mg l^{-1} in the Lake Prášilské. The nonlabile monomeric Al was only about 1% of the total reactive Al in the lakes Černé and Čertovo. Notice that the lakes Černé and Prášilské differ in Al concentrations, not in pH. For comparison, two waters from the study area are included in Fig. 2: the Lake Laka and a brownwater brook near Kvilda. Concentrations of DOC in the lakes Černé, Čertovo, Prášilské and Laka were 2.2, 2.5, 4.7 and 6.3 mg l^{-1}, respectively (analyses from August 1986).

In a group of acidified, metal-contaminated lakes of Sudbury (Canada), studied by MacIsaac *et al.*(1987), there are some comparable with ours with respect to low pH and high Al concentrations. Their most acid lakes were inhabited by only a few rotifer species, most often by *Keratella taurocephala* (a species not present in Europe) and *Gastropus* sp. *Polyarthra remata* was present in their lakes, but not typically in the acidic ones. Planktonic Crustacea were found over the whole pH spectrum in acidified lakes of Sudbury (Yan & Strus, 1980) and La Cloche Mountains (Sprules, 1975). Many species, however, were eliminated in low pH until in some lakes a single species remained. Replacement of crustaceans by a rotifer *Brachionus urceolaris* was described by Havas & Hutchinson (1983) in shallow, naturally acidified arctic ponds, in conditions of very extreme acidity and toxicity. We have found this species in the Šumava lakes either, but only occasionally and in very low densities. Its abundance in acid waters seems to be determined primarily by food.

Carter *et al.* (1986) found distinct zooplankton with reduced species richness in weakly buffered lakes of Atlantic Canada, which they explained by competition, rather than by any toxic effect. The concentrations of Al in their lakes were, however, less than 0.43 mg l^{-1}. Havens & DeCosta (1986) observed decline of zooplankton in their acid- and Al-treated enclosures, but it was difficult to decide between direct toxic effects and indirect ones. The Al concentrations were less than 0.3 mg l^{-1}.

With respect to concentrations of the two Al fractions the Lake Prášilské is similar to the most degraded, fishless lake of Hornström *et al.* (1984), inhabited by *Bosmina coregoni*, *Asplanchna priodonta* and *Polyarthra remata*.

Toxic effects of other metals than Al in the three lakes under study cannot be excluded. Veselý (1985) gives data on total Be, Cd, Zn, Pb and Cu from the three lakes. In the lakes Čertovo and Černé, cadmium reached concentrations classified as 'very high' (> 0.3 $\mu g\ l^{-1}$ according to the Swedish Environmental Protection Agency, 1991). Later on, Veselý *et al.* (1989) studied speciation of Al and Be in the lakes Čertovo and Prášilské and suggested that a part of toxicity of acid waters which have been attributed to aluminium, may actually be caused by beryllium.

Frič & Vávra (1897) quoted brown trout (*Salmo trutta*) as the only fish inhabiting the lakes Černé and Čertovo before 1890. From 1890 the brook trout (*Salvelinus fontinalis*) was repeatedly introduced into the Lake Černé. The population died off in the mid-1970's. Now the three lakes are apparently void of fish.

Lakes in the High Tatra Mountains

Our previous studies of the Tatra lakes (Stuchlík *et al.*, 1985) revealed important changes in concentration of the maior anions. Sulphate and nitrate increased several times while bicarbonate decreased by 100 $\mu eq\ l^{-1}$ on the average. Of 132 lakes and small standing waters, 84% had ANC less than 100 $\mu eq\ l^{-1}$. It was shown that NO_x, besides SO_2, contributes significantly to acidification of waters in the region, as it follows from the high concentrations of nitrate in the lakes (31% of the sum $SO_4^* + NO_3$). The sensitivity of a lake to acidification and its stage in the acidification process is determined by the sum Ca + Mg, but Mg was less than 10% of the sum in the lakes under study (Fott *et al.*, 1987, Fott *et al.*, 1992; Kopáček & Stuchlík, 1994).

The next step was confrontation of the acidification status of a lake, as determined by the chemical factors, with the aquatic life. Distribution of Crustacea in the lakes was studied earlier

Hydrobiologia **274**: 49–56, 1994.
J. Fott (ed.), Limnology of Mountain Lakes.

Chemical characteristics of lakes in the High Tatra Mountains, Slovakia

Jiří Kopáček[1] & Evžen Stuchlík[2]
[1] *Hydrobiological Institute, Czech Acad. Sci., Na sádkách 7, 370 05 České Budějovice, Czech Republic;*
[2] *Department of Hydrobiology, Charles University, Viničná 7, 120 44 Praha 2, Czech Republic*

Key words: acidification, mountain lakes, High Tatra Mountains

Abstract

The chemistry of 53 lakes at various stages of acidification and inhabited (at the presence and/or in the past) by pelagic Crustacea was studied in September 1984. Ten of these lakes were investigated in detail biannually (July and October 1987–1990). The July results reflect the influence of snowmelt and were compared with the October ones. The most important anion was sulphate with the average values of 98 and 104 $\mu eq\,l^{-1}$ in 1984 and 1987–1990, respectively. High concentrations of nitrate (21–56 $\mu eq\,l^{-1}$) were observed in lakes above the treeline. Mean relative composition of cations does not differ between July and October; small changes are in the mean relative composition of anions. Acidification of lakes, expressed as a decrease in alkalinity, is 100 $\mu eq\,l^{-1}$, and is equal to the increase in the sum of sulphate and nitrate. The values of total phosphorus and COD are the lowest in the range of pH 5–6.5. Alkalinity, sulphate, nitrate and pH do not show any trend with time over the last ten years.

Introduction

Acid deposition has resulted in acidification of freshwaters in large regions of Europe and North America characterized by geology that is resistant to chemical weathering (Likens & Bormann, 1974; Wright & Gjessing, 1976; Wright & Henriksen, 1978; Wright *et al.*, 1980). The High Tatra Mountains (the Carpathian system; 20° 10′ E, 49° 10′ N; the highest peak reaches 2655 m) are situated on such geologically sensitive area. Their central part is composed of granite with calcareous inclusions in several parts. There are about 120 lakes of glacial origin in the Slovak part of the mountains, and these are mostly situated in alpine zone (1800–2200 m a.s.l.). This area receives acid deposition (annual weighted mean of pH is 4.2–4.5) with sulphate and nitrate the dominant anions (Cerovský, 1983; Gazda & Lopašovský, 1983; Moldan *et al.*, 1987).

It is difficult to specify the onset of acidification of this region by using only chemical data from the lakes, despite the fact that many data from the past are available (Olszewski, 1939; Stangenberg, 1938; Juriš *et al.*, 1965; Bombówna, 1965; Ertl, 1965; Oleksynowa, 1970), Records of zooplankton, however, indicate that acidification occurred in the years 1975–1978 (Stuchlík *et al.*, 1985; Fott *et al.*, 1987). Actually 42% of the lakes were acidified and the alkalinity of 86% of lakes was less than 100 $\mu eq\,l^{-1}$ in the years 1981–1983 (Stuchlík *et al.*, 1985; Fott *et al.*,1987; Fott *et al.*, 1992).

The purpose of this paper is to present data on the chemical composition and acidification of lakes in the Slovak part of the High Tatra Mountains obtained in the years 1984 to 1990.

Materials and methods

53 lakes selected on the basis of a previous synoptic survey were sampled in September 1984. The altitude of the lakes range from 1350 to 2150 m a.s.l., the area from 0.1 to 20 ha, and the maximum depth from 0.5 to 54 m. All of these lakes are and/or were in the past populated by zooplankton (Fott *et al.*, 1987; Fott *et al.*, 1994). During the years 1987–1990, ten representative lakes (Table 1) were investigated according to the recommendations of Programming manual of International cooperative programme for assessment and monitoring of acidification of rivers and lakes. The beginning of July and October each year was chosen as a sampling period. The July data reflect the chemical situation during or just after snowmelt, while October is the period of minimum variability of the lake water chemical composition.

Samples of water were collected from the surface layer at the deepest point of lake, prefiltered through a 40 μm polyamide mesh to remove zooplankton, and kept in plastic bottles in the dark at 4 °C.

Alkalinity (Gran titration, Mackereth *et al.*, 1978) and pH were measured after overnight storage of the samples at 20 °C. Major ions from the survey in September 1984 were determined immediately in our field laboratory in the High Tatra Mountains with exclusion of magnesium, sodium and potassium which were determined by atomic absorption spectroscopy (AAS) in the laboratory of Czech Geological Survey in Prague. Calcium was analysed using ionic selective electrode Orion. Nitrate was determined spectrophotometrically after reduction to nitrite by means of spongy cadmium, chloride by Gran argentometric titration with silver ionic selective electrode Orion and sulphate as sum of the strong acids after subtraction of nitrate and chloride (Mackereth *et al.*, 1978).

During the survey conducted in 1987–1990 ammonium was determined by the rubazoic acid method (Procházková, 1964) immediately after samples were brought to the field laboratory. Chemical oxygen demand (COD) was analysed within 2–5 days according to Hejzlar & Kopáček (1990). Samples for determination of cations by AAS were preserved with HNO_3, samples for total phosphorus (determination according to Popovský, 1970) were preserved with H_2SO_4. Nitrate was determined spectrophotometrically after reduction to nitrite by alkaline hydrazine (Procházková, 1959), chloride was analysed also spectrophotometrically according to Zall *et al.* (1956). Sulphate was determined using the technique of capillary isotachophoresis (Zelenský *et al.*, 1984).

Table 1. Main physical characteristics of lakes studied from 1987 to 1990.

Lake	Type	Altitude ma.s.l.	Area ha	Max. depth m	Significant in/outflow	Watershed
ŠTRBSKÉ	Eutroficated	1346	19.8	20.0	–/–	Forest
POPRADSKÉ		1494	6.7	17.6	+/+	Forest, rocks, meadows
MALÉ HINCOVO	Non-acidified	1923	2.2	6.4	–/+	Meadows, rocks
VEL'KÉ HINCOVO		1946	20.1	53.7	+/+	Rocks, meadows
L'ADOVÉ		2057	1.7	17.8	–/–	Rocks
BATIZOVSKÉ	Acidified	1879	3.5	8.7	+/+	Rocks, meadows
VYŠNÉ WAHLENBERGOVO		2145	5.2	20.0	–/–	Rocks
SLAVKOVSKÉ	Strongly acid	1676	0.1	2.5	–/–	Dwarf pine
STAROLESNIANSKÉ		2000	0.7	4.2	–/+	Meadows
JAMSKÉ	Dystrophic	1447	0.7	4.2	+/–	Forest

Results and discussion

Chemistry of 53 lakes

The concentrations of ions in 53 lakes sampled in September 1984 are presented in Table 2. Mean values were calculated for groups of lakes with the same range of alkalinity, that is for non-acidified lakes (alkalinity higher than 25 μeq l^{-1}), acidified lakes (alkalinity 25–0 μeq l^{-1}) and strongly acidified lakes (alkalinity less than 0 μeq l^{-1}). Appropriate values of pH or calcium could be used instead of alkalinity to distinguish these three main groups of lakes. Concentrations of other cations do not significantly differ within the groups.

The most important anion is sulphate which ranges from 61 to 160 μeq l^{-1}. The mean concentration of nitrate is 37 μeq l^{-1}. This is more than in the most of the acidified areas in the world (Henriksen & Brakke, 1988; Eilers *et al.*, 1988a; Eilers *et al.*, 1988b; Brakke *et al.*, 1988; Wathne *et al.*, 1989; Mosello *et al.*, 1991). Concentrations of both sulphate and nitrate have increased nearly 10 times since 1937 (data from Stangenberg, 1938), and they do not vary significantly within lake group. The concentrations of calcium span a wide range from 42 to 358 μeq l^{-1}.

All lakes are affected by acid deposition but the degree of acidification depends on the concentration of calcium or calcium plus magnesium and on preacidification alkalinity, respectively (Henriksen, 1979; Henriksen, 1980). The least-squares linear regression between the actual concentrations of bicarbonate (μeq l^{-1}) and calcium or calcium plus magnesium (μeq l^{-1}) for the 53 lakes gave:

$$[HCO_3^-] = 0.87[Ca] - 76.8 \qquad R^2 = 0.90;$$

$$[HCO_3^-] = 0.80[Ca + Mg] - 78.8 \qquad R^2 = 0.94.$$

The lowest and highest conductivity (25 °C) was 8 and 54 μS cm^{-1}. Average value was 26 μS cm^{-1}.

There are only slight differences in the relative composition of cations between non-acidified and acidified lakes (Fig. 1). An essential portion of hydrogen (13%) appears in strongly acidified lakes. On the other hand there are significant changes in relative composition of anions in all three categories of lakes.

Chemistry of 10 lakes, July and October 1987–1990

The mean chemical parameters of 10 selected lakes sampled from 1987 to 1990 calculated for both July and October are presented in Table 3. Concentrations of alkalinity, sulphate, nitrate and all cations except hydrogen and ammonium were significantly lower in July than in October, while concentrations of chloride did not change. The mean autumn increase in concentrations of major cations is 10–15%, alkalinity 20%, sulphate and nitrate 10%. High concentrations of nitrate were found in lakes above the treeline; the mean October value in Batizovské lake was 56 μeq l^{-1}. Nitrate was very low in lakes with forested watersheds (Štrbské, Jamské and Slavkovské lakes).

The ratio of sulphate to chloride ranged from 5 to 17 with the average value of 11.6 and 12.7 in

Table 2. Chemical characteristics of 53 lakes (September 1984). Mean values (μeq l^{-1}) are calculated for 3 categories of lakes with the same range of alkalinity.

Type of lake	Number	pH	H^+	Ca^{2+}	Mg^{2+}	Na^+	K^+	NO_3^-	SO_4^{2-}	HCO_3^-	Σ^+	Σ^-
Non-acidified (Alk > 25)	32	6.6	0	202	19	17	4	37	101	100	242	238
Acidified (0 < Alk < 25)	13	5.8	2	114	9	10	3	42	87	10	138	139
Strongly acidified (alk < 0)	8	4.8	16	74	13	14	5	31	107	0	122	138

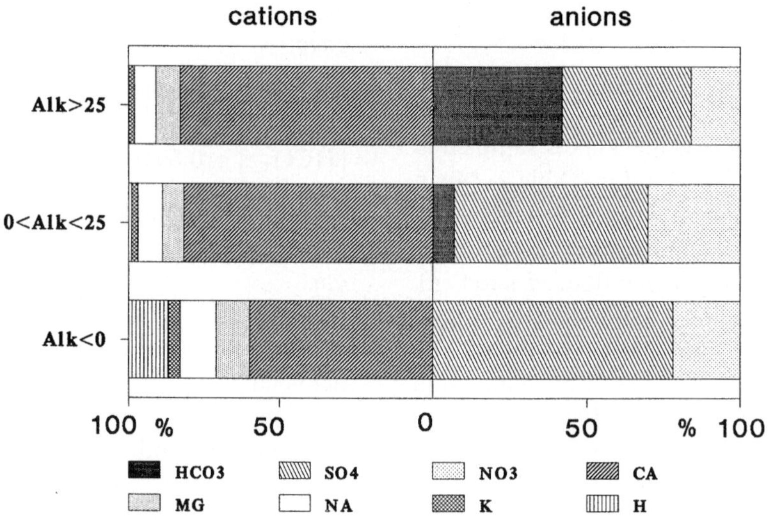

Fig. 1. The relative ionic composition in three categories of lakes investigated in September 1984.

July and October, respectively. The ratio of calcium to magnesium was constant during the year and in lakes with small watersheds (Jamské, Štrbské, Starolesnianské, Slavkovské) close to

Table 3. Chemical characteristics of 10 lakes (1987-1990). Mean July (J) and October (O) values (μeq l^{-1}).

Lake	Month	pH	Ca^{2+}	Mg^{2+}	Na^{+}	K^{+}	NH_4^{+}	NO_3^{-}	SO_4^{2-}	Cl^{-}	HCO_3^{-}	Σ^{+}	Σ^{-}
ŠTRBSKÉ	J	6.9	183.0	84.5	30.8	12.0	1.4	0.2	130.0	25.0	129.0	311.8	284.2
	O	7.0	178.0	86.7	32.0	12.0	0.9	0.2	144.0	24.0	134.0	309.7	302.2
POPRADSKÉ	J	6.6	157.0	15.3	16.8	3.5	0.6	41.5	92.0	9.0	50.0	193.4	192.5
	O	6.8	202.0	20.4	22.9	5.3	0.8	54.0	100.0	9.0	92.0	251.5	255.0
MALÉ HINCOVO	J	7.2	320.0	88.5	17.5	5.6	1.3	32.0	149.0	7.5	244.0	433.0	432.5
	O	7.3	340.0	92.5	19.7	6.6	1.7	33.0	164.5	8.1	255.0	460.6	460.6
VEL'KÉ HINCOVO	J	6.8	197.0	17.5	15.0	4.3	0.5	42.0	92.0	6.7	92.0	234.5	232.7
	O	6.8	199.0	18.0	16.4	5.0	0.6	44.5	96.0	6.0	92.0	239.1	238.5
L'ADOVÉ	J	5.6	88.0	8.0	9.5	3.3	9.0	42.0	67.0	8.0	5.5	120.3	122.5
	O	6.5	154.0	10.0	13.8	1.6	0.5	45.0	70.0	7.0	59.0	180.2	181.0
BATIZOVSKÉ	J	5.9	117.0	9.6	15.0	3.4	0.8	50.3	84.5	6.5	4.5	147.1	145.8
	O	6.2	142.0	11.7	19.6	4.0	0.4	56.0	97.0	5.8	21.2	178.3	180.0
VYŠNÉ	J	5.0	76.0	8.5	7.5	2.0	8.5	44.5	76.0	8.5	0.0	111.6	129.0
WAHLENBERGOVO	O	5.3	94.0	10.3	9.8	3.3	0.8	46.0	75.5	6.3	0.0	123.1	127.8
SLAVKOVSKÉ	J	4.7	38.5	14.0	19.5	7.5	0.7	0.2	102.0	7.3	0.0	101.1	109.5
	O	4.8	38.5	15.0	22.0	9.0	1.2	1.2	104.0	8.9	0.0	101.5	114.1
STAROLESNIANSKÉ	J	4.5	37.5	9.0	8.3	4.0	4.5	21.5	75.0	6.5	0.0	94.2	103.0
	O	4.8	53.5	12.0	11.1	3.7	3.0	21.0	96.7	6.5	0.0	100.7	124.2
JAMSKÉ	J	4.5	60.0	18.3	25.5	6.3	0.2	0.5	132.0	13.5	0.0	142.7	146.0
	O	4.5	60.0	19.5	26.3	6.2	0.3	2.0	128.0	14.5	0.0	143.2	144.5
Mean	J	5.8	127.4	27.3	16.5	5.2	2.8	27.5	100.0	9.9	52.5	189.0	189.8
	O	6.0	146.1	29.6	19.4	5.7	1.0	30.3	107.6	9.6	65.3	208.8	212.8

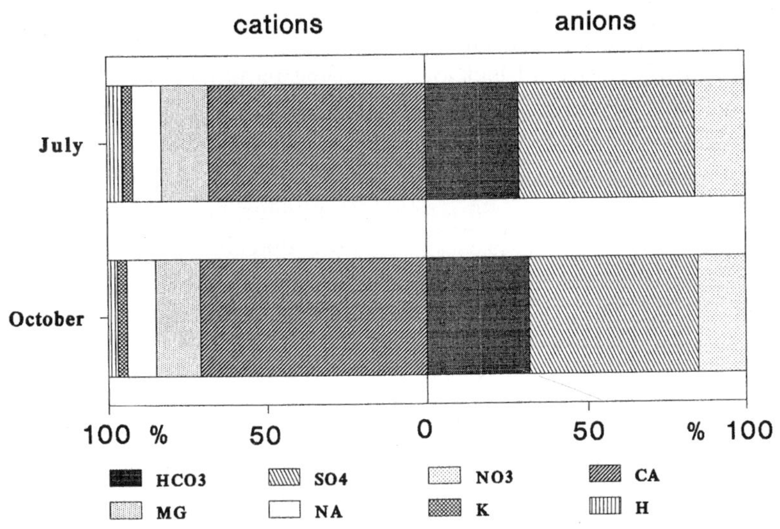

Fig. 2. The relative ionic composition in 10 lakes investigated in July and October 1987–1991.

the ratio in deposition (3.3–4.3; Cerovský, 1983, Gazda & Lopašovský, 1983). The ratio ranged between 10 and 15 in lakes with larger watersheds.

The differences between sum of cations and anions (Table 3) are caused by the presence of aluminium in strongly acidified or dystrophic lakes and organic anions in lakes with the forested watersheds. Neither aluminium nor organic anions are included in ionic balance.

Mean July concentrations of total aluminium in strongly acidified lakes and dystrophic lakes were higher than in October and ranged from 70 μg l^{-1} (Vyšné Wahlenbergovo in October 1987) to 440 μg l^{-1} (Slavkovské in July 1989). Total aluminium was usually lower than 20 μg l^{-1} in non-acidified lakes.

The relative composition of cations calculated as an average for all 10 selected lakes is similar in both July and October (Fig. 2), while the mean relative concentrations of anions differs slightly. Sulphate decreases from 52.5% to 50.5%, and bicarbonate increases from 28.0% to 30.5%. Higher differences were found in lakes strongly affected by snowmelt; sulphate decreases from 48% to 39% in Popradské lake and from 55% to 39% in L'adové lake, and bicarbonate increases from 25% to 36% in Popradské lake and from 4% to 35% in L'adové lake.

Mean October values of COD, TP and absorbance at 254 nm are plotted against pH in Fig. 3. The points fit Almer's relation for interactions within dissolved organic matter, phosphorus, alu-

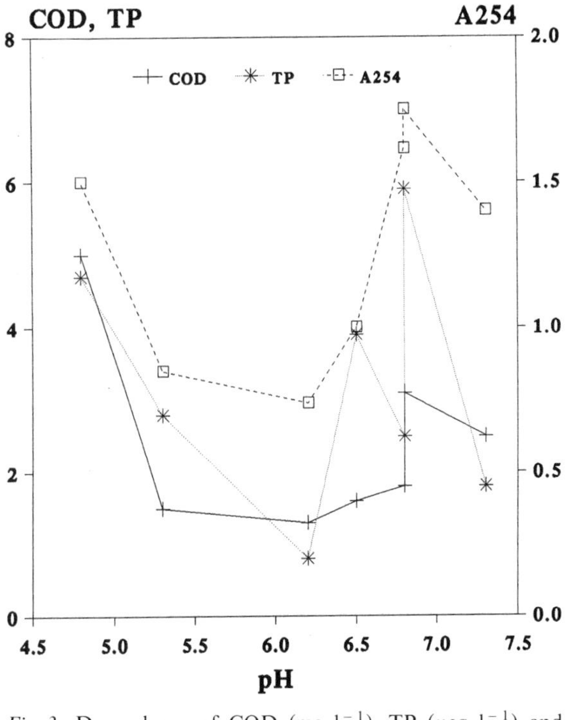

Fig. 3. Dependence of COD (mg l^{-1}), TP (μeq l^{-1}) and absorbance at 254 nm (A m^{-1}) on pH. Mean October values in 7 selected lakes from Table 3.

minium and pH (Almer *et al.*, 1978) and reach a minimum at pH between 5 and 6.5. Extremely low concentrations of chlorophyll-*a* and lack of zooplankton in lakes with pH between 5 and 6.5 (Fott *et al.*, 1994; Vyhnálek *et al.*, 1994) are probably caused by this decrease in concentration of phosphorus.

Long-term trends

October pH of all 10 selected lakes do not show any trend over time during the period 1981–1990, and the fluctuation of values during one season often exceeds differences between the years.

Mean October values of sulphate, alkalinity and nitrate calculated for 5 selected lakes (Popradské, Vel'ké Hincovo, L'adové, Batizovské, Vyšné Wahlenbergovo) from 1981 to 1990 are presented in Fig. 4. Sulphate seems to drop slightly in last 3 years, possibly due to the effect of dry years. Nitrate and alkalinity do not show any significant trends.

Acidification of lakes

Some parameters of acidification calculated according to Wright (1983) such as the difference between pre-acidification alkalinity and measured concentrations of bicarbonate plus hydrogen and aluminium are presented in Table 4. We assume here that concentrations of both calcium and magnesium do not change during acidification. Sea-salt correction is based on chloride (Henriksen, 1980).

Resulting acidification level is 99 μeq l^{-1} for 53 lakes in 1984 and 100 and 102 μeq l^{-1} for 10 lakes (1987–1990) in July and October, respectively. The contribution of nitrate to acidificaion is more than 30%.

We assume that the calculated acidification is somewhat overestimated because of high concentrations of non-marine calcium in deposition; 22 μeq l^{-1} in rain (Cerovský, 1933) and 43 μeq l^{-1} in snow-water (Gazda, Lopašovský, 1983). Some portion of strong acids calculated in the

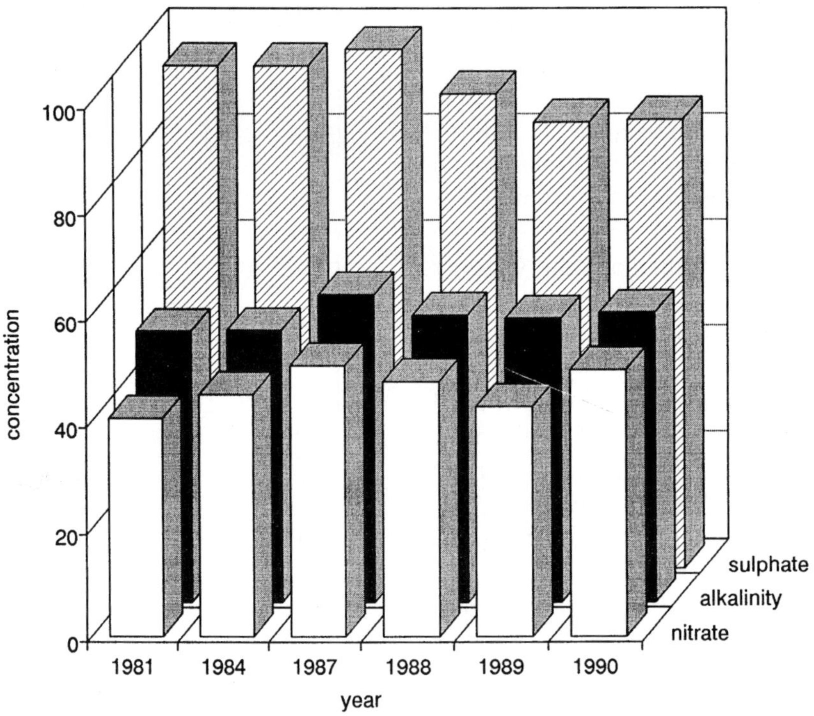

Fig. 4. Long-term trends of sulphate, nitrate and alkalinity. Mean October concentrations (μeq l^{-1}) for Popradské, Velké Hincovo, L'adové, Batizovské and Vyšné Wahlenbergovo lakes in 1981–1990.

Table 4. Acidification of the High Tatra Mountains lakes calculated according to Wright (1983). All concentrations in μeq l^{-1}.

$bSO_4^* = 0.09(Ca^* + Mg^*) + Na^* + K^*$ $nSO_4^* = SO_4 - bSO_4$		$dAlk = 0.91(Ca^* + Mg^*) - HCO_3 + H + Al$ $Al = 10^{(12.7 - 2.56pH)}$	
	1984 (53 lakes)	1987-1990 (10 lakes)	
	September	July	October
bSO_4^*	26	27	32
nSO_4^*	71	72	74
$nSO_4^* + NO_3$	109	100	105
dAlk	99	100	102
		$dAlk = Slope \times (nSO_4^* + NO_3)$	
Slope	0.91	0.99	0.97

acidification model as net $SO_4^{2-} + NO_3^-$ value is in fact neutralized directly in the atmosphere; on the other hand the acidification calculated from sum of $Ca^{2+} + Mg^{2+}$ values is overestimated because of the portion of calcium from deposition.

No correlation between acidification and altitude was found perhaps because there is only a 400 m difference in altitude of lakes situated above the treeline (1600 m a.s.l.). On the other hand the level of acidification depends on size of lake and its watershed. Small and shallow lakes with small watersheds are more easily affected by snowmelt and rainwater than larger lakes with long residence times and large watersheds.

Conclusions

(1) All lakes in the Slovak part of the High Tatra Mountains are affected by acidification. Mean decrease in alkalinity (about 100 μeq l^{-1}) is equal to the mean increase of sulphate (about 70 μeq l^{-1}) plus nitrate (about 30 μeq l^{-1}). The vulnerability of lakes to acidification and the degree of acidification of the lakes is determined by the concentration of calcium.

(2) The melting period is very important in the process of acidification. The pH values drop drastically both in the lakes with short residence time and in the lakes with low alkalinity. The July sums of ions are lower than the October ones. Mean relative composition of cations (Ca = 70%, Mg = 15%, Na = 9%, K = 3%, H = 4%) is nearly the same in both July and October. Sulphate decreases and alkalinity increases from July to October.

(3) Significant drop in total phosphorus, COD and absorbance (254 nm) was recorded within the lakes with pH 5–6.5.

(4) Data of pH, alkalinity, sulphate and nitrate do not show any significant trend of acidification from 1981 to 1990.

References

Almer, B. W., W. Dickson, C. Ekström & E. Hörnström, 1978. Sulfur pollution and the aquatic ecosystem. In J. O. Nriagu (ed.), Sulfur in the environment, part II: Ecological impacts. Wiley, New York: 271–311.

Bombówna, M., 1965. Hydrochemical investigations of the Morskie oko lake and the Czarny staw lake above the Morskie oko in the Tatra Mountains. Kom. Zagosp. Ziem. Gorsk. PAN 11: 7–17.

Brakke, F. D., D. H. Landers & J. M. Eilers., 1988. Chemical and physical characteristics of lakes in the Northeastern United States. Envir. Sci. Technol. 22: 155–163.

Cerovský, M., 1983. Meranie znečistenia ovzdušia na Chopku a zrážok v Bratislave. In Źborník referátov zo IV. celoštátneho hydrogeochemického seminára Hydrogeochemické problémy znečist'ovania prírodných vôd. GÚDŠ, Bratislava: 73–80. [Measurement of air pollution at Mt. Chopok and pollution of precipitation in Bratislava].

Eilers, J. M., D. F. Brakke & D. H. Landers, 1988a. Chemical and physical characteristics of lakes in the Upper Midwest, United States. Envir. Sci. Technol. 22: 164–171.

Eilers, J. M., D. H. Landers & D. F. Brakke, 1988b. Chemical and physical characteristics of lakes in the Southeastern United States. Envir. Sci. Technol. 22: 172–177.

Ertl, M. 1965. Chemizmus Popradského plesa. (In Slovak with English summary). Sborník prác o Tatranskom národnom parku 8: 45–55. [Chemism of lake Poprad].

Fott, J., E. Stuchlik & Z. Stuchlíková, 1987. Acidification of the lakes in Czechoslovakia. In B. Moldan & T. Pačes (eds), Extended Abstracts of International workshop on Geochemistry and Monitoring in Representative Basins (GEOMON), Prague, Czechoslovakia, April 27–May 1, Prague: 77–79.

Fott, J., E. Stuchlík, Z. Stuchlíková, V. Straškrabová, J. Kopáček & K. Šimek, 1992. Acidification of lakes in the Tatra Mountains (Czechoslovakia) and its ecological consequences. In R. Mosello, B. M. Wathne & G. Giussani (eds), Limnology on groups of remote mountain lakes: ongoing and planned activities. Documenta Ist. ital. Idrobiol. 32: 69–81.

Fott, J., M. Pražáková, E. Stuchlík & Z. Stuchlíková, 1994. Acidification of lakes in Šumava (Bohemia) and in the High Tatra Mountains (Slovakia). Hydrobiologia 274/Dev. Hydrobiol. 93: 37–47.

Gazda, S. & K. Lopašovský, 1983. Chemické zloženie zimných zrážok na území Slovenska. In Zborník referátov zo IV. celoštátneho hydrogeochemického seminára Hydrogeochemické problémy znečist'ovania prírodných vôd. GÚDŠ, Bratislava: 63–71. [Chemical composition of winter precipitation in Slovakia].

Hejzlar, J. & J. Kopáček, 1990. Determination of low chemical oxygen demand values in water by the dichromate semimicro method. Analyst 115: 1463–1467.

Henriksen, A., 1979. A simple approach for identifying and measuring acidification of freshwater. Nature 278: 542–545.

Henriksen, A., 1980, Acidification of freshwaters – a large scale titration. Proc., Int. Conf. Ecol. Impact Acid. Precip., SNSF-project, Norway: 68–74.

Henriksen, A. & D. F. Brakke, 1988. Increasing contributions of nitrogen to the acidity of surface waters in Norway. Wat. Air Soil Pollut. 42: 183–201.

Juriš, S., M. Ertl, E. Ertlová & M. Vranovský, 1965. Niektoré poznatky z hydrobiologického výzkumu Popradského plesa. Zborník prác o Tatranskom národnom parku 8: 33–34. [Some remarks resulting from the hydrobiological research of Popradské lake].

Likens, G. E. & F. H. Borman, 1974. Acid rain: a serious regional environmental problem. Science 184: 1176–1179.

Mackereth. F. J. H., J. Heron & J. F. Talling, 1978. Water analysis: Some revised methods for limnologists. FBA Scientific Publication No 36, 120 pp.

Moldan, B., M. Vesely & A. Bartoňová, 1987. Chemical composition of atmospheric precipitation in Czechoslovakia, 1976-1984-I. Monthly samples. Atmosph. envir. 21: 2383–2395.

Mosello, R., A. Marchetto, G. A. Tartari, M. Bovio & P. Castello, 1991. Chemistry of alpine lakes in Aosta valley (N. Italy) in relation to watershed characteristics and acid deposition. Ambio 20: 7–12.

Oleksynowa, K., 1970. Geochemical characterization of the waters in the Tatra Mountains. Acta Hydrobiol. 12: 1–110.

Olszewski, P., 1939. Kilka danych o chemizmie wod w okolici Hali Gasienicowej. Einige bestimmungen zum chemismus der gewasser in der umbegung der Gasienicowa-Alm (Hohe Tatra). Sprawozdania Komisji Fizjograficznej Polskiej Akademii Umiejetnisci 72: 501–530.

Popovský, J., 1970. Determination of total phosphorus in fresh waters. Int. Revue ges. Hydrobiol. 55: 435–443.

Procházková, L., 1959. Bestimmung der Nitrate im Wasser. Z. analyt. Chem. 167: 254–260.

Procházková, L., 1964. Spectrophotometric determination of ammonia as rubazoic acid with bispyrazolone reagent. Analyt. Chem. 36: 865–871.

Stangenberg, M., 1938. Zur Hydrochemie der Tatraseen. Verh. int. Ver. Limnol. 8: 211–220.

Stuchlík, E., Z. Stuchlíková, J. Fott, L. Růžička & J. Vrba, 1985. Vliv kyselých srážek na vody na území Tatranského národního parku. (In Czech with English summary). Zborník prác o Tatranskom národnom parku 26: 173–211. [Effect of acid precipitations on waters of the TANAP territory].

Vyhnálek V., J. Fott & J. Kopáček, 1994. Chlorophyll–phosphorus relationship in acidified lakes of the High Tatra Mountains (Slovakia). Hydrobiologia 274/Dev. Hydrobiol. 93: 171–177.

Wathne, B. M., R. Mosello, A. Henriksen & A. Marchetto, 1989. Comparison of the chemical characteristics of mountain lakes in Norway and Italy. In M. Johannessen, R. Mosello & H. Barth (eds), Acidification processes in remote mountain lakes. Air pollution research report 20: 41–58.

Wright, R. F., 1983. Predicting acidification of North American lakes. Acid Rain Res. Rept. 4/83. Norwegian Institute for Water Research, Oslo, 165 pp.

Wright, R. F. & E. T. Gjessing, 1976. Changes in the chemical composition of lakes. Ambio 5: 219–223.

Wright, R. F. & A. Henriksen, 1978. Chemistry of small norwegian lakes, with special reference to acid precipitation. Limnol. Oceanagr. 23: 487–498.

Wright, R. F., W. Conroy, T. Dickson, R. Harriman, A. Henriksen & L. Schofield, 1980. Acidified lake district of the world: a comparison of water chemistry of lakes in southern Norway, southern Sweden, southwestern Scotland, the Adirondack Mountains of New York, and southeastern Ontario. Proc., Int. Conf. Ecol. Impact Acid. Precip., SNSF-project, Norway: 377–379.

Zall, D. M., D. Fisher & M. Q. Garner, 1956. Photometric determination of chlorides in water. Analyt. Chem. 28: 1665–1668.

Zelenský I., V. Zelenská, D. Kanianski, P. Havaši & V. Lednárová, 1984. Determination of inorganic anions in river water by column-coupling capillary isotachophoresis. J. Chromatogr. 294: 317–323.

Hydrobiologia 274: 57–64, 1994.
J. Fott (ed.), Limnology of Mountain Lakes.

Paleolimnological records of carotenoids and carbonaceous particles in sediments of some lakes in Southern Alps

Andrea Lami [1], Aldo Marchetto [1], Piero Guilizzoni [1], Anselma Giorgis [1] & Julieta Masaferro [2]
[1] *C.N.R. Istituto Italiano di Idrobiologia, 28048 Verbania-Pallanza;* [2] *PROGEBA-CONICET Appartado N. 47, 8400 S.C. de Bariloche, Provincia de Rio Negro, Argentina*

Key words: sediment, core stratigraphy, algal and bacterial development, acidification

Abstract

Stratigraphic analyses of organic carbon, organic nitrogen and algal and bacterial carotenoids in short cores of profundal sediments of four alpine lakes (Tovel, Leit, Paione superiore and Tom) were used to reconstruct their trophic history. In addition, depth distribution of carbonaceous particle concentrations provided information on lake contamination from atmospheric deposition.

In three lakes (Tovel, Leit and Tom), sedimentary carotenoids unique to sulfur photosynthetic bacteria (okenone and isorenieratene) provide evidence of changes in the oxygen, light and sulfide conditions in the water column. All the lakes are oligotrophic or moderately productive, and the algal community is dominated by Chlorophyta, Pyrrhophyta and Cryptophyta. Cyanobacteria are rather poorly represented.

The steep increase of carbonaceous particles in the uppermost sediment layers of all the lakes suggests that lake contamination by atmospheric transport of pollutants began in the 1940s to 1950s. These data, coupled with those from a parallel study on Chrysophycean scale-inferred pH, indicate recent acidification in those which are poorly buffered (Paione superiore and Leit).

Introduction

This study was primarily designed to consider factors related to lake acidification in remote areas, as more than 50% of some 320 studied lakes in the Alps are susceptible to acidification (Mosello, 1984). In some of these lakes the pH during the past has been inferred by analyzing assemblages of chrysophytes (Marchetto & Lami, 1994) and by a pigment index (Guilizzoni *et al.*, 1992a).

However, it seemed appropriate first of all to look at this specific problem in a wider context which also took into account the trophic characteristics of the selected lakes. Carbon, nitrogen and algal pigments were therefore analyzed in order to describe trophic and phytoplankton (and perhaps also bacterial) history during the past century. Moreover, trends in atmospheric deposition of substances deriving from the combustion of fossil fuels have been traced from soot particles counted in sediment cores from four alpine lakes.

Materials and methods

Sediment cores were collected in two Italian (Paione superiore and Tovel) and Swiss (Tom and Leit) Alpine lakes. Lakewater pH ranged from 5.7 to 8.0. Lake Tovel and Lake Tom are characterized by having the highest pH and being located in a watershed rich in carbonates (up to 80–90% dry sediment); Lake Tom also shows a high sulphate concentration. On the other hand, lakes Paione superiore and Leit are poorly buffered (range of alkalinity of 5–22 μeq l^{-1}). Some selected geographical, physical and chemical characteristics of the study lakes are reported in Marchetto & Lami (1994) and in Guilizzoni *et al.* (1992b).

Short sediment cores (up to 22 cm) were collected using a gravity corer from the deepest point of the lakes (Leit = 12 m; Paione superiore = 8 m; Tom = 4 m; Tovel = 38 m). Cores were immediately sliced at 1 cm intervals on the lake shore.

The main technique used for dating the sediments was ^{210}Pb although ^{137}Cs analysis was also useful at some sites. Lake Tovel shows the highest sedimentation rate (0.2–0.25 cm yr^{-1}), whereas for lakes Leit, Tom and Paione superiore it ranged between 0.10 and 0.15 cm yr^{-1}.

Analysis of carbonaceous particles deriving from fossil fuel combustion was carried out following the methods of Renberg & Wik (1985) and Rose (1990). This technique involves oxidation of the sample with nitric acid to remove organic matter, and sedimentation of the resulting suspension on the floor of a glass Petri dish. After evaporation of the supernatant, the black spherules are counted using a stereomicroscope at × 40 magnification. A scanning electron microscope (SEM) was also employed in some cases.

Algal pigments were extracted from the sediment with 90% acetone, and total carotenoids were quantified using the equation proposed by Züllig (1982). Specific algal and bacterial carotenoids were analyzed by High Performance Liquid Chromatography (HPLC) following the method of Mantoura & Llewellyn (1983) with some modifications.

Organic carbon and organic nitrogen were analyzed using a CHN analyzer (Carlo Erba).

Results and discussion

Organic carbon and organic nitrogen

Concentration curves for C and N are shown in Fig. 1. The most evident increase in the uppermost layer is observed in Lake Tovel (from 4 to

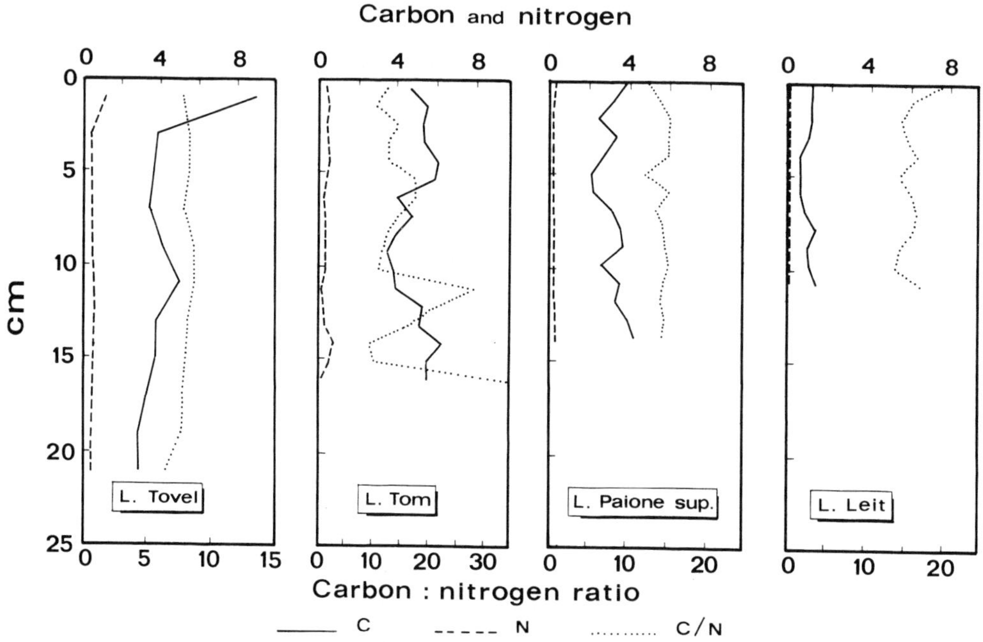

Fig. 1. Organic carbon, nitrogen and C:N ratio distribution in sediment cores of four Alpine lakes (% dry wt^{-1}).

9% d.w.). With the exception of Lake Leit, the values along the cores for all the lakes are very similar and related to the organic matter content. These values, although lower, are of the same order of magnitude as those measured on the River Po plain (e.g. Guilizzoni *et al.*, 1982). Lake Tom shows the largest fluctuation throughout the core length. The C and N values in Lake Leit (range for C = 0.51–1.33% d.w.; for N = 0.03–0.08% d.w.) are lower.

On the basis of the natural variation of the C/N ratio, and in accordance with our expectations, the sedimented organic matter in three of the lakes would appear to be prevalently of allochthonous origin (values higher than 10; Kemp, 1971). Only Lake Tovel has C/N values below 10, which are constant along the core depth.

Algal and bacterial carotenoids

Pigment concentrations could be used as indirect measures of the photosynthetic standing crop and algal community composition (Züllig, 1981; Sanger, 1988). Total carotenoids, and the ubiquitous chlorophyll *a* and β-carotene, can be used as measures of the quantity of total algal community. Lutein, fucoxanthin, alloxanthin, diadinoxanthin, zeaxanthin and dinoxanthin are used as markers of specific taxonomic groups within the algal community (Züllig, 1982). Two other carotenoids (okenone and β-isorenieratene) found in the sediments of lakes Tovel, Tom and Leit are known as markers of some anaerobic bacteria belonging to the purple and green phototrophic forms (Brown *et al.*, 1984; Züllig, 1985, 1986).

Total carotenoid depth profiles in the sediments are similar to those of β-carotene and Chl. *a* with relatively high concentrations in Lake Leit and Lake Tovel (Figs 2, 3). Lake Tom shows a single high value of about 800 μg g^{-1} organic weight at 5 cm of core depth. In this lake as well as in Lake Tovel it was also possible to detect the Chl. *b*. Sedimentary pheophytin *a* was always lower than chlorophylls in lakes Tom and Tovel. Analyses of per cent native chlorophyll (not shown), an index of pigment preservation (Swain, 1985), show high values (up to 70%) only in these two lakes. In addition, the ratio between the spectrophotometric reading at 430 nm and 410 nm of the acetone extract, an other index of pigment pres-

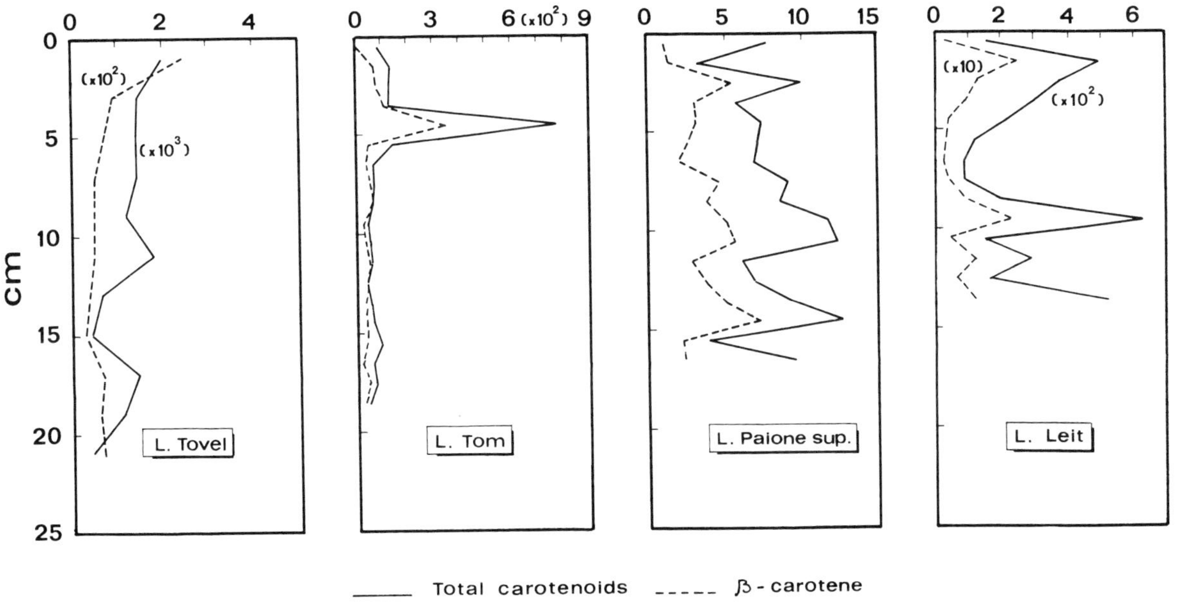

Fig. 2. Depth profiles of total carotenoids and β-carotene (μg g^{-1} organic matter) in four Alpine lakes.

Relationships between water chemistry, geographical and lithological features of the watershed of Alpine lakes located in NW Italy. Verh. int. Ver. Limnol. 24: 155–157.

Moss, B., 1967. A spectrophotometric method for the estimation of percentage degradation of chlorophylls to pheopigments in extracts of algae. Limnol. Oceanogr. 12: 335–340.

Paganelli, A., A. Miola & P. Cordella, 1988. Il Lago di Tovel (Trentino) e la circolazione delle sue acque. Riv. Idrobiol. 27: 363–376.

Renberg, I. & M. Wik, 1985. Soot particle counting in recent lake sediments. An indirect dating method. Ecological Bulletins 37: 53–57.

Rose, N. L., 1990. A method for the extraction of carbonaceous particles from lake sediment. J. Paleolimnol. 3: 45–53.

Sanger, J. E., 1988. Fossil pigments in paleoecology and paleolimnology. Palaeogeogr. Palaeoclim. Palaeoecol. 62: 343–359.

Swain, E. B., 1985. Measurement and interpretation of sedimentary pigments. Freshwat. Biol. 15: 53–75.

Züllig, H., 1982. Untersuchungen über die Stratigraphie von Carotinoiden im geschichteten Sediment von 10 Schweizer Seen zur Erkundung früherer Phytoplankton-Entfaltungen. Schweiz. Z. Hydrol. 44: 1–98.

Züllig, H., 1985. Pigmente phototropher Bakterien in Seesedimenten und ihre Bedeutung für die Seenforschung. Schweiz. Z. Hydrol. 47: 87–126.

Züllig, H., 1986. Carotenoids from plankton and photosynthetic bacteria in sediments as indicators of trophic changes in Lake Lobsigen during the last 14000 years. Hydrobiologia 143/Dev. Hydrobiol. 37: 315–319.

Hydrobiologia **274**: 65–74, 1994.
J. Fott (ed.), Limnology of Mountain Lakes.

Algal flora of lakes in the High Tatra Mountains (Slovakia)

Jaromír Lukavský
Academy of Sciences of Czech Republic, Institute of Botany, Section of Ecology, Dukelská 145, 379 82, Třeboň, Czech Republic

Key words: acidification, altitude, algal flora, mountain lakes, species richness, the High Tatra Mountains

Abstract

Eighty seven from a total of 120 lakes in the Slovak part of the High Tatra Mts. have been visited since 1982. Their summer phytoplankton and algae growing on stones were collected, identified and documented.

Some species that are interesting, rare, or not previously known in Slovakia have been found: *Clastidium setigerum*, *Colacium calvum*, *Chroococcus subnudus*, *Chr. quaternarius*, *Coelastrum printzii*, *Coenocystis quadriguloides*, *Oocystis naegelii*, *Scopulonema polonicum*, *Thelesphaera alpina*, *Trochiscia prescottii*.

The number of algal species, found in the open water of the lakes, decrease with altitude and increase with pH.

Introduction

The first mentions of algae from the High Tatra Mts. come from the mid-18th century, when prospectors believed that coloured snow indicated places where ore veins rose close to the surface. Phycologists have visited this area since the middle of the 19th century. Unfortunately, the majority of papers published till now have focused on those groups of algae (diatoms, desmids or dinoflagellates) which can be easily collected by a net, preserved, and studied later. We have a great shortage of data concerning small phytoplankton which must be studied alive. Another shortcoming is that most phycologists paid attention only to their favourite groups of algae and did not mention other genera. That is why we have only fragmentary knowledge of the algal flora of the lakes.

A remarkable exception is a paper published by Juriš & Kováčik (1978) which really focused on the whole algal flora of the lakes; they studied 44 lakes, line algae were collected using only a plankton net. In addition, the unique paper of Starmach (1973) deals with very important benthic algae of lakes in the H. Tatra Mts. A bibliography on the algal flora of the H. Tatra Mts. was published by Lhotský (1971).

My main objective was a taxonomical determination of algae found in the H. Tatra lakes in order to evaluate the present state with respect of acidification of the lakes (Stuchlík *et al.*, 1985; Fott *et al.*, 1987). I have focused my attention on sensitive species which are not evaluated in paleolimnological studies. Besides determination and documentation of the algal species I tried to establish whether the number of species found in a lake (species richness) is related to the lake's altitude and acidification status.

Description of the sites studied

There are about 120 lakes in the Slovak part of the H. Tatra Mts. I have visited 87 of them, some of them several times, since 1982. The lakes are of glacial origin, most of them are oligotrophic with a low content of both organic and inorganic compounds. Principal rock of the studied area is granite. For more detailed information see Stuchlík *et al.* (1985); Fott *et al.* (1994); Kopáček & Stuchlík (1994).

A few lakes are dystrophic (Jamské, Rakytovské, Slepé, Štrbské, Trojrohé) other ones are eutrophicated (Popradské, Štrbské). The largest lake on the Slovak side of the H. Tatra Mts., is Velké Hincovo with content almost a half million cubic meters of water. Dozens of 'lakes' are only temporary, shallow pools, filled with water from melting snow only during the spring.

Material and methods

Samples of 2 to 5 litres were collected from the surface and concentrated at the site by pressure filtration through membrane filters (Synpor 3–4, Barvy laky Co., porosity 0.6–0.85 μm) into volumes ca 10 ml. The concentrated samples were transferred to the laboratory and centrifugated at 1500 rpm using a small desktop centrifuge. The seston concentrated into a drop was used for microscopical examination. This technique allows the concentration of live algae from very dilute suspensions. When both the filtration and centrifugation steps are performed gently, even sensitive species like *Cryptomonas* and *Chromulina* remain alive.

Algae growing in shallow water on stones were collected by scratching with use of a toothbrush and also examined alive.

As a simple measure of species richness, I calculated average numbers of species found in individual lakes. The arithmetic means (per a visit) were calculated from data of Juriš & Kováčik (1987), B. Fott (unpublished) and of mine. Hydrological and hydrochemical data are from Stuchlík *et al.* (1985) and Tržilová & Miklošovičová (1989).

Results and discussion

Algal flora of the lakes

A detailed study of the algal flora of the lakes is under preparation. Some comments on rare algae or on species not previously known in Slovakia are presented here:

Binuclearia tectorum (Kütz.) Beger in Wichmm. 1937. (Plate IV. Fig. 10). This alga was described as *Binuclearia tatrana* Wittr. 1886 from Štrbské lake, H. Tatra Mts. It is a very conspicuous alga, easy to recognize by the H-shaped rests of mother cell walls. Filaments of the alga are often released into plankton. It is not rare in cold, dystrophic lakes, and in peat bogs and wet slopes of the northern hemisphere (Lukavský, 1970). H. Tatra Mts.: in Štrbské and Jamské lake.

Botryococcus pila Komárek & Marvan 1992. (Plate IV., Fig. 14). This species was identified and described from material collected by the present author in Slepé lake a small puddle in a peatbog by Štrbské lake. It was floating in a dense layer on the surface of the water. Fossil findings of the species are common in peaty deposits under name *Pila*, from Late Glacial and postglacial period.

Chroococcus quaternarius Zalessky (Plate I, Fig. 18), (Syn: *Gloeocapsa turgida f. quaternaria* (Zalessky) Hollerb.). This species, new for Slovakia, was found in benthos of Malé Černé lake.

Chroococcus subnudus (Hansg.) Cromb. & Kom. (Plate I., Fig. 2), (Syn: *Ch. turgidus v. subnudus* Hansg., *Gloeocapsa turgida f. subnuda* (Hansg.) Hollerb.). A large species, similar to *Ch. turgidus*. According to Cronberg & Komárek (in press) it is a characteristic species in metaphyton and secondary in plankton of purely oligotrophic or slightly dystrophic swamps and lakes. Known from the northern parts of the temperate zone and from mountains in central Europe. In the H. Tatra Mts.: in benthos of the small, dystrophic Rakytovské Nižné lake.

Clastidium setigerum Kirchn. 1880 (Plate I., Fig. 8). I have found it on stones in the littoral of Litvorové lake and Žabie Javorové lake. It was mentioned by Hansgirg 1892, as an epibiont on

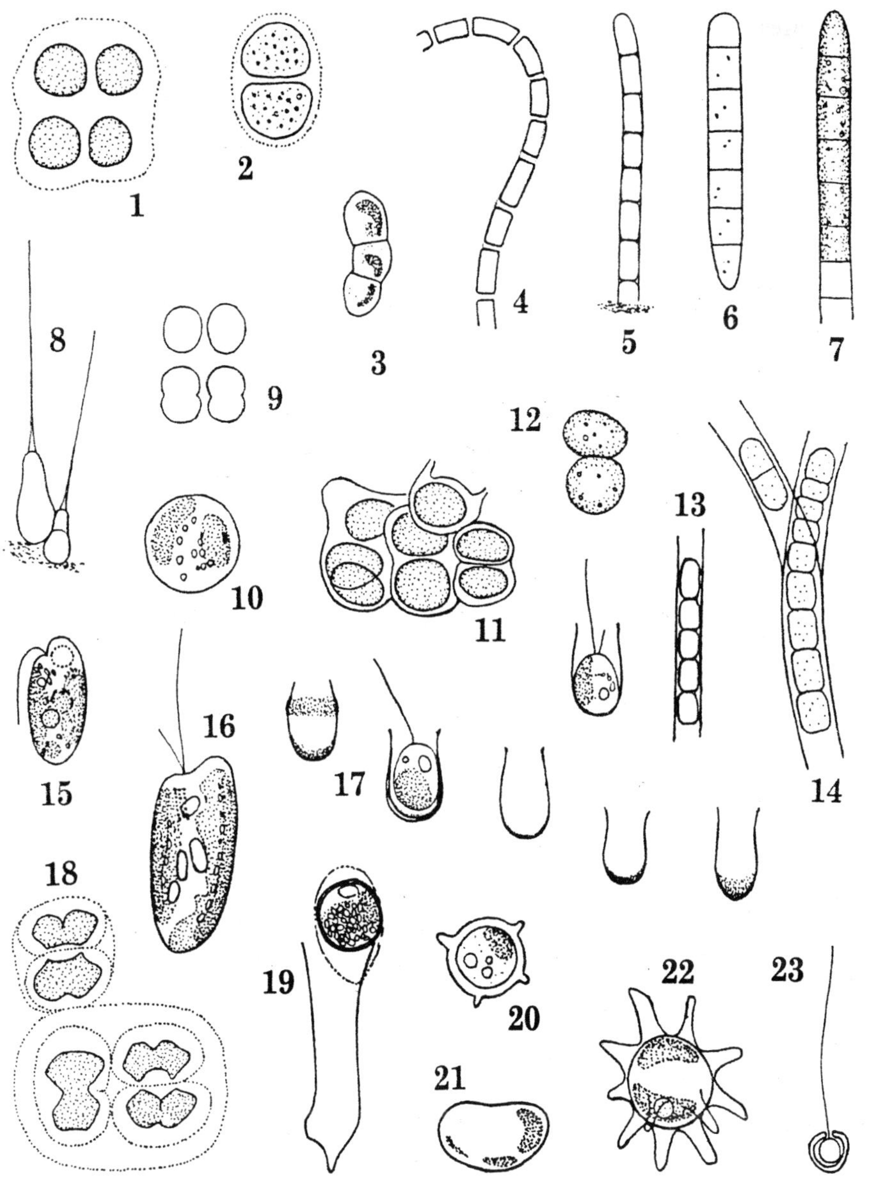

Plate I. Fig. 1 – *Merismopedia cf. glauca* (Ehrenb.) Näg. [Rakytovské nižné]; 2 – *Chroococcus subnudus* (Hansg.) [Rakytovské nižné]; 3 – Hormogonia [Pusté, Velká studená dolina]; 4 – *Pseudanabaena catenata* Lauternb. [Rakytovské nižné], 5 – *Heteroleibleinia cf. fontana* (Kütz. & Hansg.) Anagn. et Kom. [Litvorové]; 6 – Hormogonium [Ladové]; 7 – *Phormidium innundatum* Kütz. [Rakytovské nižné]; 8 – *Clastidium setigerum* Kirchn. [Litvorové]; 9 – *Merismopedia cf. elegans* A. Braun [Trojrohé]; 10 – *Chromulina sp.* [Velké černé]; 11 – *Scopulonema polonicum* (Racib). Geitler [Litvorové]; 12 – *Synechocystis sp.* [Biele]; 13 – *Leptolynbyia lurida* Anagn. et Kom. [Batizovské]; 14 – *Pseudophormidium cf. tenue* (Thur.) Forty. [Batizovské]; 15 – *Cryptomonas obovata* Skuja [Biele]; 16 – *Cryptomonas erosa* Ehrenb. [Slepé]; 17 – *Pseudokephyrion tatricum* (Juriš) Starm. / syn: *Kephyriopsis tatrica* Juriš. / [a,c,d,-Litvorové, b- Zelené Krivánské, e,f – Malé Hincovo]; 18 – *Chroococcus quaternarius* Zalessky [Malé černé]; 19 – *Dinobryon pediforme* (Lemm.) Stein [Trojrohé]; 20,22 – Cyst of *Chrysophyceae* [Zmrzlé Bielovodské, Štrbské]; 21 – Cyst of *Dinophyceae* [Zmrzlé Bielovodské]; 23 – *Chrysococcus rufescens* Klebs. [Pusté].

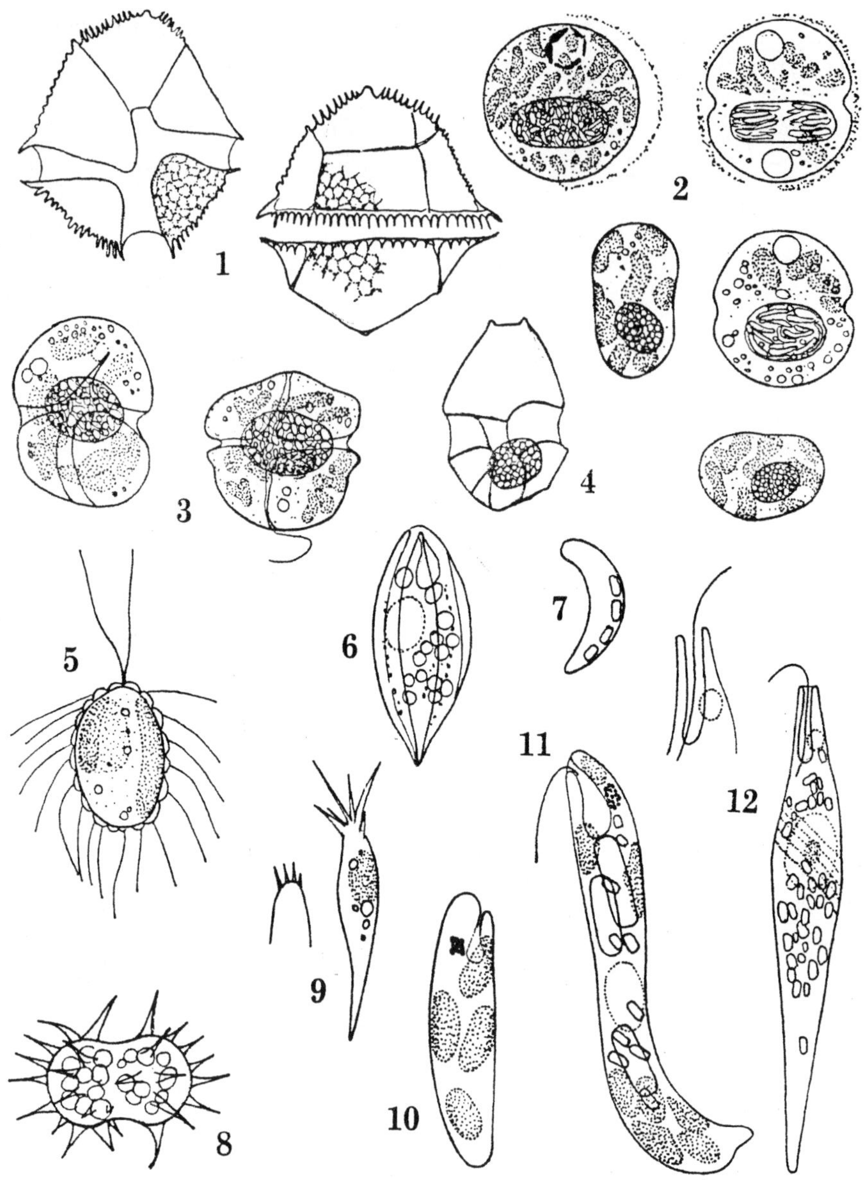

Plate II. Fig. 1 – *Peridinium cf. willei* Huiff.-Kass. [Štrbské]; 2 – *Gymnodinium sp.* [Zelené Krivánské]; 3 – *Gymnodinium uberrimum* (Allman) Kofoid & Swezy [Štrbské, Jamské]; 4 – *Peridinium sp.* [Zelené krivánské]; 5 – *Mallomonas robusta* Matv. [Velické]; 6 – *Petalomonas sp.* [Malé černé]; 7 – *Menoidium sp.* [Štrbské]; 8 – Cyst of *Dinophyceae* [Velké Hincovo]; 9 – *Mallomonas acrocomos* Ruttn. [Velké černé, Okrúhlé]; 10 – *Colacium calvum* Stein / syn: *Euglena physeter* Fott [Štrbské]; 11 – *Euglena ehrenbergii* Klebs [Biele]; 12 – *Astasia klebsii* Lemm. [Štrbské].

Cladophora, surprisingly in fishponds and springs near Prague. The nearest recent locality outside Slovakia is the Alps (Geitler, 1977).

Coelastrum printzii Rayss 1915. (Plate V., Fig. 5). (Syn: *Sorastrum simplex* Wille 1879, *Coelastrum humicola* Gistl 1935). In small, dystrophic lakes in Canada, Norway, Novaya Zemlja. H. Tatra Mts.: in Kriváňské Zelené lake.

Coenocystis quadriguloides Fott 1974 (Plate III., Fig. 5): The species was identified in dystrophic waters of the H. Tatra Mts. Now in Malé Hincovo lake.

Colacium calvum Stein, 1878 (syn. *Euglena physeter* Fott (Plate II., Fig. 10). The species was identified by B. Fott from fishponds. It occurred in Štrbské lake (slightly dystrophic and eutrophicated). Later it has been recognized as a motile stage of the species *Colacium calvum* Stein by Willey (1982).

Oocystis naegelii A. Br. 1855. (Plate IV., Fig. 5)

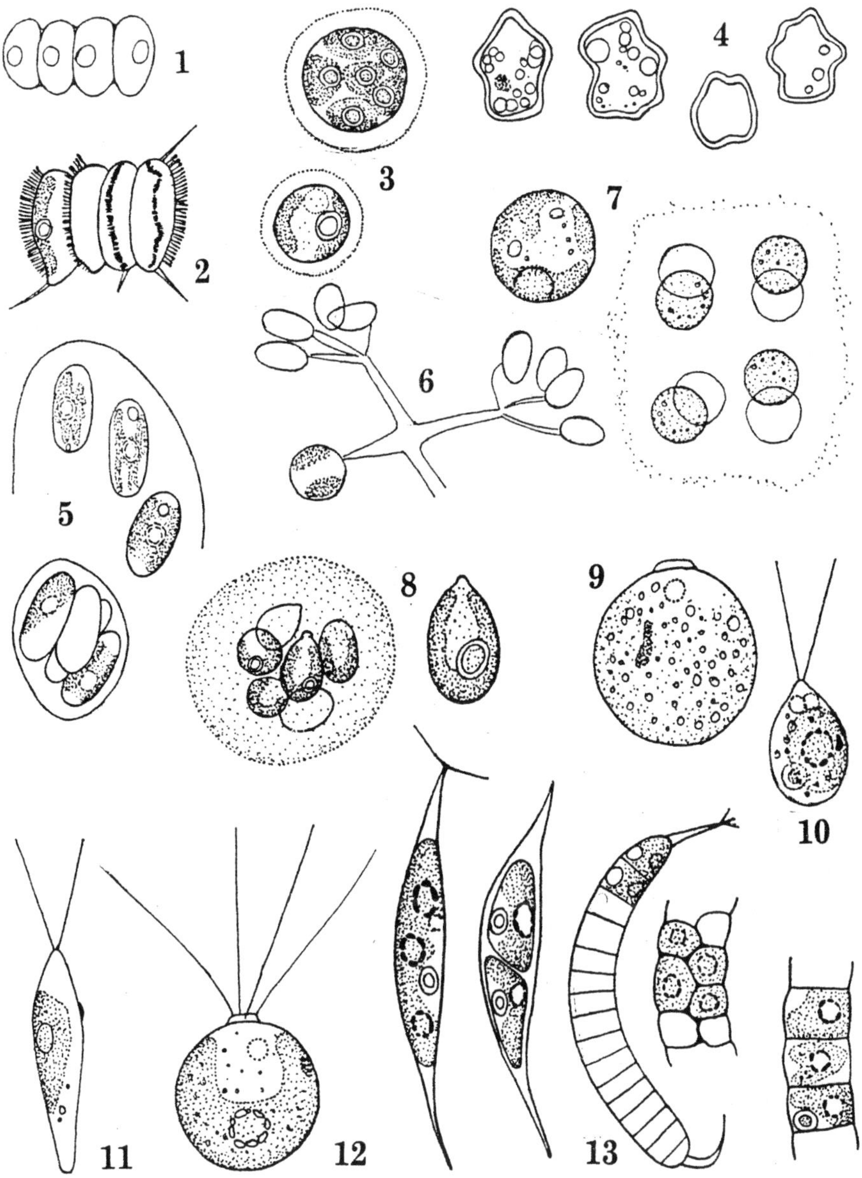

Plate III. Fig. 1 – *Scenedesmus ecornis* (Ralfs) Chod. [Biele]; 2 – *Scenedesmus helveticus* Chod. [Popradské]; 3 – *Planktosphaeria gelatinosa* G. M. Smith [Okrúhlé, Velké Hincovo]; 4 – cf. *Thelesphaera alpina* Pasch. or propagation bodies of *Hepaticae* [Velké Hincovo]; 5 – *Coenocystis quadriguloides* Fott [Malé Hincovo]; 6 – *Dictyosphaerium tetrachotomum* Printz [Jamské]; 7 – *Eutetramorus fottii* (Hind.) Kom./syn: *Coenococcus fottii* Hind./[Štrbské]; 8 – *Sphaerellocystis ellipsoidea* Ettl [Štrbské]; 9 – *Chloromonas sp.* [Starolesnianské]; 10 – *Chlamydomonas sp.* [Starolesnianské]; 11 – *Chlamydomonas isogama* Korsch. in Pasch. [Biele]; 12 – *Carteria multifilis* (Fresenius) Dill. [Slepé]; 13 – *Korshikoviella limnetica* (Lemm.) Silva [Tiché, Nižné Furkotské, on *Daphnia pulicaria*].

(Syn: *O. nordstedtiana* (De-Toni) Playf.: Known from small water bodies at mountains in Scandinavia and Central Europe. H. Tatra Mts: In the dystrophic Trojrohé lake.

Pseudokephyrion tatricum (Juriš) Starmach 1985. (Syn. *Kephyriopsis tatrica* Juriš) (Plate I., Fig. 17): This small flagellate with a fine lorica was described by Juriš (1964) from Popradské

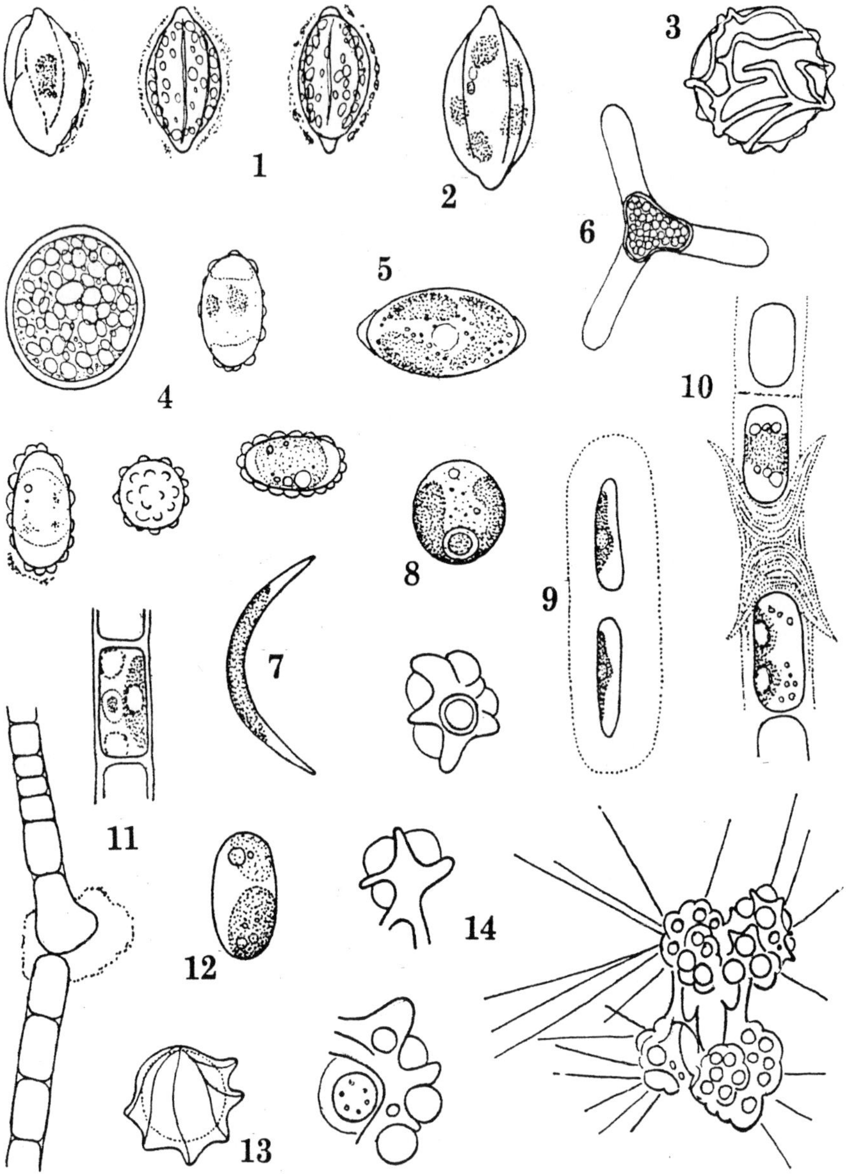

Plate IV. Fig. 1, 2 – *Scotiella nivalis* (Shuttlew) Fritsch / = *Chloromonas cryophila* Hoham & Mullet/ [snow field close Zelené kačacie, Okrúhlé]; 3 – *Trochiscia prescottii* Siemin [Rakytovské nižné]; 4 – *Chlamydomonas rostafinskii* Starmach & Kawecka, zygotes [snow field by Zelené kačacie]; 5 – *Oocystis naegelii* A. Braun [Trojrohé]; 6 – *Chionaster nivalis* (Bohl.) Wille [Zmrzlé Bielovodské, Okrúhlé, snow field by Zelené kačacie]; 7 – *Monoraphidium cf. arcuatum* (Korsch.) Hind. [Velké Hincovo]; 8 – *Chlorella vulgaris* Beij. [Popradské]; 9 – *Elakatothrix biplex* (Nyg.) Hind. [Štrbské]; 10 – *Binuclearia tectorum* (Kütz.) Beger in Wichmm. [Malé černé, Jamské]; 11 – *Ulothrix sp.* [Zelené kačacie]; 12 – *Trachychloron cf. depauperatum* Pasch. [Wahlenbergovo vyšné]; 13 – *Coelastrella striolata* CHOD. [[Rakytovské nižné]; 14 – *Botryococcus pila* Komárek & Marvan [Štrbské, Slepé].

lake. This lake, previously oligotrophic, has been eutrophicated. *Pseudokephyrion tatrica*, however, occurs also in other lakes (Litvorové, Zelené Kriváňské, Malé Hincovo, Starolesnianské, Nižné Žabie Bielovodské). It shows a wide variability in the structure of lorica.

Scotiella cf. nivalis (Shuttlew) Fritsch (= akinets of *Chloromonas cryophila* Hoham & Mullet 1977) (Plate IV., Fig. 1) It is a cryoseston species, common in permanent snow fields (red snow), from which it is washed down into some lakes.

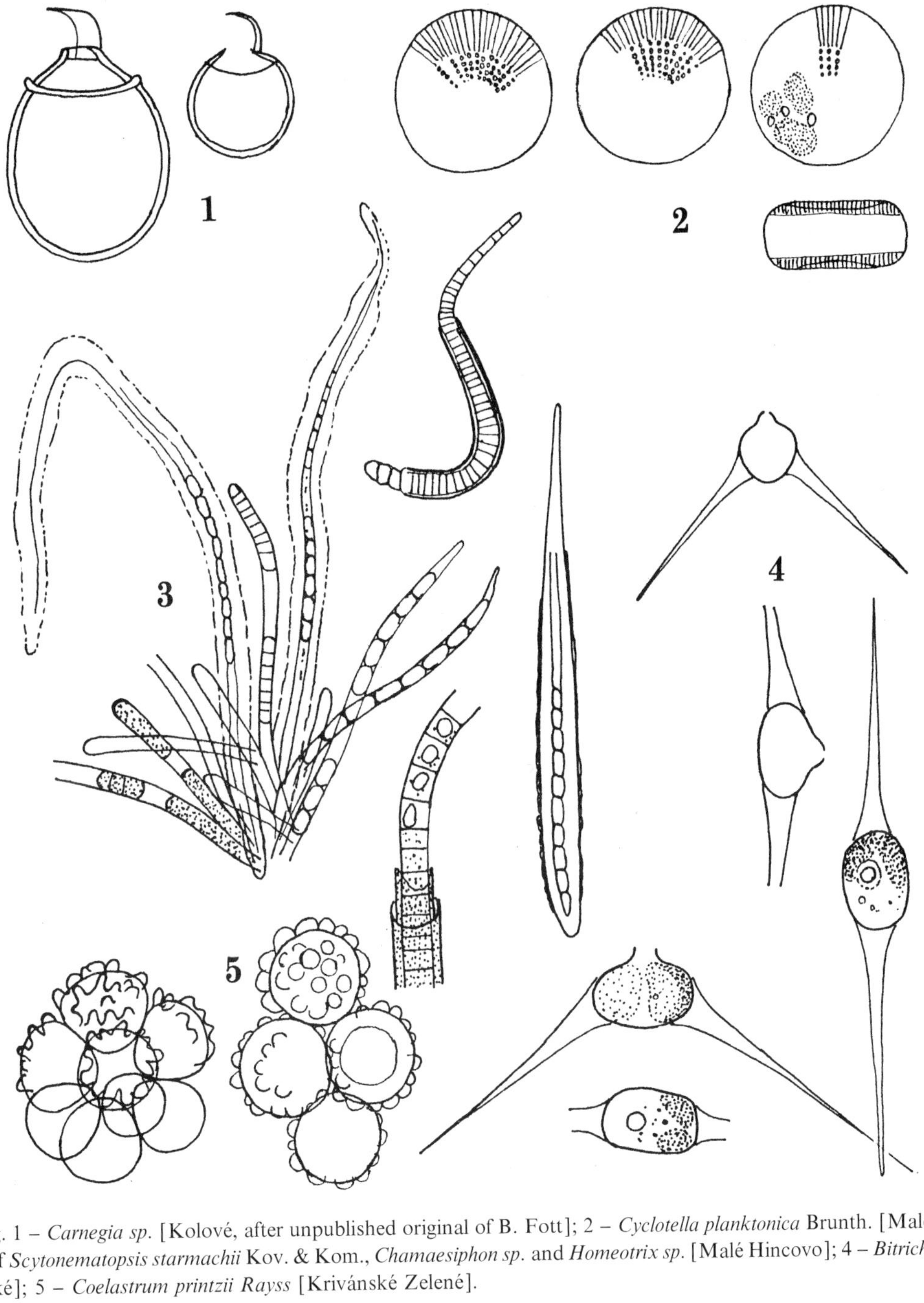

Plate V. Fig. 1 – *Carnegia sp.* [Kolové, after unpublished original of B. Fott]; 2 – *Cyclotella planktonica* Brunth. [Malé Hincovo]; 3 – cluster of *Scytonematopsis starmachii* Kov. & Kom., *Chamaesiphon sp.* and *Homeotrix sp.* [Malé Hincovo]; 4 – *Bitrichia phaseolus* Fott [Jamské]; 5 – *Coelastrum printzii Rayss* [Kriváнské Zelené].

The H. Tatra Mts: Okrúhlé, Zelené Kačacie and other lakes.

Scytonematopsis starmachii Kováčik & Komárek 1987 (Plate V. Fig. 3). Filaments with false branching, adhering to rocky substrate by a rounded ensheathed base. The blue-green alga shows a great variability, e.g. *Dichothrix-*, *Ammatoidea-*, *Scytonema*-like filaments and branching. Heterocyts are only occasional. This alga is widely distributed in lakes and 'Spritzzone' on granite substrates (Kováčik & Komárek, 1988). Filaments are released into plankton, too. The H. Tatra Mts.: Malé Hincovo, Slavkovské lake.

cf. *Thelesphaera alpina* Pasch. 1943 (Plate III., Fig. 4). According to Hindák (p.c.) they are only propagation bodies of Hepaticae, e.g. *Lophozia*. It will be necessary to observe the reproduction of the organism. The nearest locality of *T. alpina* is the Alps. H. Tatra Mts: in hypolimnion of V. Hincovo lake.

cf. *Trochiscia prescottii* Sieminska 1965 (Plate IV., Fig. 3): The nearest locality is Canada. Unfortunately, I have not yet seen reproduction. The H. Tatra Mts: in Rakytovské Nižné lake.

Species richness of phytoplankton in the High Tatra lakes

The lakes included in the analysis are situated at altitudes ranging from 1200 to 2300 m above sea level. A plot of phytoplankton species richness (number of algal species found in the open water) against altitude revealed a tendency of decreasing

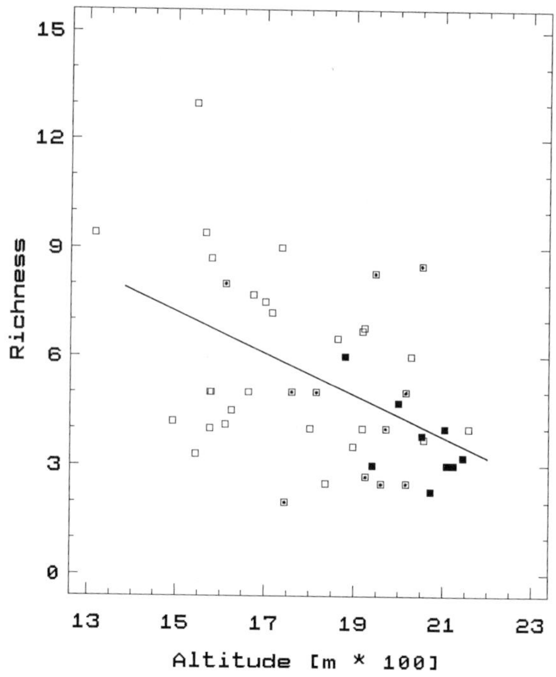

Fig. 1. Relation of species richness (number of algal species found in the open water of a lake) to altitude. The estimated line of simple linear regression ($r = -0.44^{**}$, $n = 49$) is drawn into the graph. The acidified lakes (alkalinity < 0, pH < 5) are labeled by ■, the threatened lakes (alkalinity 0–25 μeq l^{-1}) by ⊡, the other lakes by □.

Table 1. Spearman and Kendall rank correlations. Values r significant at 0.05 are labelled by (*), significant at 0.01 by (**). Correlation coefficient (sample size, significance level).

Spearman rank correlations (above the diagonal); Kendall rank correlations (below the diagonal)

	Altitude	Sum. temp.	Volume	pH	Alkal.	Richness
Altitude	–	**−0.16** (41,0.30)	**0.06** (43,0.68)	**−0.23** (41,0.15)	**−0.49*** (31,0)	**−0.46**** (50,0)
Sum. temp.	**−0.12** (64,0.26)	–	**−0.27** (41,0.08)	**−0.35*** (41,0.02)	**−0.40** (22,0.07)	**0.13** (34,0.45)
Volume	**0.039** (64,0.71)	**−0.18** (42,0.91)	–	**0.56**** (41,0)	**0.21** (22,0.33)	**0.28** (35,0.1)
pH	**−0.16** (64,0.12)	**−0.24*** (42,0.02)	**0.39**** (43,0)	–	**0.62**** (22,0)	**0.38*** (34,0.027)
Alkal.	**−0.32**** (64,0.01)	**−0.29*** (42,0.05)	**0.15** (43,0.31)	**0.48**** (41,0)	–	**0.09** (28,0.64)
Richness	**−0.32**** (64,0)	**0.06** (42,0.59)	**0.21** (43,0.07)	**0.30**** (41,0)	**0.07** (35,0.56)	–

richness with altitude (Fig. 1). The graph also illustrates how lakes of different acidification status (grouping of lakes according to Stuchlík *et al.*, 1985) are distributed along the axis of altitudes: the acidified lakes lie above 1900 m, the intermediate lakes ('threatened ones') are above 1600 m.

In order to test whether species richness is a monotonically increasing (or decreasing) function of independent variables like altitude, water temperature in summer, lake volume, pH or alkalinity, Kendall and Spearman rank correlation coefficients were used; the usage of simple nonparametric tests *in lieu* of regression analysis was recommended by Sokal & Rohlf (1981). From the correlation matrix (Table 1) it follows that the number of algal species in the open water of a lake is lowest at high altitudes and in low pH. There is a clear dependence of alkalinity on altitude (see also the Fig. 1), whereas the dependence of pH on altitude is not significant. It should be emphasized that the low-pH lakes included in the statistical analysis are acidified clearwater lakes, not brownwater humic lakes.

The decrease of species richness of algae in lakes with altitude is in accordance with the findings of Gentry (1988). Sometimes, however, the relationship is reversed, or the highest diversity was found at the intermediate altitudes (Wilson *et al.*, 1990).

Acknowledgements

The work was supported by Acad. Sci. of Czech Republic, Grant # 60504. I thank also Dr J. Fott, Dr E. Stuchlík and other colleagues from Charles University in Prague and from the Institute of Hydrobiology at Č. Budějovice for some data, assistance in collecting the samples as well as friendship and enthusiasm which is necessary in the mountains. Management of The High Tatra National Park showed a great understanding for our work. Doc. Dr J. Hrbáček, and Doc. Dr J. Komárek kindly went through the manuscript.

References

Cronberg, G. & J. Komárek, in press. Review of planktic Cyanoprokaryotes found in South Swedish lakes during the 12th IAC symposium. Arch. Hydrobiol./Suppl. Algological Studies.

Fott, J., E. Stuchlík & Z. Stuchlíková, 1987. Acidification of lakes in Czechoslovakia. In B. Moldan & T. Pačes (eds), Extended Abstracts Int. Workshop. Geochemistry and Monitoring Representative Basins /GEOMON/, Prague, April 27–May 1: 77–79.

Fott, J., M. Pražáková, E. Stuchlík & Z. Stuchlíková, 1994. Acidification of lakes in Šumava (Bohemia) and in the High Tatra Mountains (Slovakia). Hydrobiologia 274/Dev. Hydrobiol. 93: 37–47.

Geitler, L., 1977. Zur Fortpflanzung der Cyanophycee *Clastidium setigerum*. Ber. Deutsch. Bot. Ges. 90: 497–499.

Gentry, A. H., 1988. Changes in plant community diversity and floristic composition on environmental and geographical gradients. Ann. Missouri Bot. Gard. 75: 1–34.

Hansgirg, A., 1892. Prodromus českých ras sladkovodních II. Arch. Přírod. výzkumu Čech, Bot. 8, 182 pp. (Prodrome of Bohemian freshwater algae II, in Czech).

Hindák, F., 1988. Contribution to the taxonomy of some cyanophyte genera. Preslia (Praha) 60: 289–308.

Hindák, F., (ed.) 1978. Sladkovodné riasy. Slovenské pedagogické nakl. Bratislava, 724 pp. (Freshwater algae, in Slovak).

Juriš, Š., 1964. *Kephyriopsis tatrica* sp. nova. Phycologia 4: 52–53.

Juriš, Š., 1976. Riasy. Zborník prác o Tatranskom národnom parku 17: 187–193. (Algae, in Slovak).

Juriš, Š. & L. Kováčik, 1987. Contribution towards the phytoplankton of lakes in the High Tatra Mts. Zbor. Slov. nár. Múz., Prír. Vedy 33: 23–40.

Komárek, J. & B. Fott, 1983. Das Phytoplankton des Süsswassers. In Huber-Pestalozzi (ed.), Die Binnengwässer 16, Stuttgart, 1044 pp.

Komárek, J. & P. Marvan, 1992. Morphological differences in natural populations of the genus *Botryococcus* (*Chlorophyceae*). Arch. Protistenkd. 141: 65–100.

Kopáček, J. & E. Stuchlík, 1994. Chemical characteristics of lakes in the High Tatra Mountains, Slovakia. Hydrobiologia 274/Dev. Hydrobiol. 93: 49–56.

Kováčik, Ľ. & J. Komárek, 1988. Scytonematopsis starmachii, a new cyanophyte species from the High Tatra Mts. (Czechoslovakia). Arch. Hydrobiol. Suppl. 80, 1–4, Algolog. Stud. 50–53: 303–314.

Lhotský, O., 1971. Bibliografia algologiczna Tatr. Acta hydrobiol. 13: 477–490. (Phycological bibliography of the Tatra Mts., in Polish).

Lukavský, J., 1970. Morphological variability and reproduction of the alga *Binuclearia tectorum* under natural conditions. Nova Hedwigia 19: 189–199.

Psenner, R., 1989. Chemistry of high mountain lakes in sili-

Table 2. Estimate of the number of lakes sensitive to acidification.

Threshold	No.	%	95% confidence interval	
			No.	%
T.A. ≤200 μeq l^{-1}	142	73%	138–147	71–76%
T.A. ≤50 μeq l^{-1}	92	47%	62–101	32–52%
H ≤6.0	35	18%	18–36	9–18%
pH ≤5.3	11	6%	1–30	0.5–15%

bility to acidification (e.g. Goldstein & Gherini, 1984). This value may ensure that in spring total alkalinity values do not fall below zero. Recently a threshold of 50 μeq l^{-1} has been proposed (e.g. Sullivan *et al.*, 1990) and this value is closer to the loss of alkalinity calculated for Alpine lakes in this and other studies (Mosello *et al.*, 1988; Psenner *et al.*, 1988).

The numbers of lakes with total alkalinity values of below 200 and 50 μeq l^{-1} are reported in Table 2 with their 95% confidence interval, showing that about two thirds of the lakes considered can be regarded as sensitive to acidification.

Conditions of acute acidification are reached when the lake pH drops below 5.3 and the acid-base equilibrium no longer depends on the buffer effect of bicarbonate ions, but on the various forms of aluminium (Wright, 1983). In these lakes the toxic effects are induced both by low pH and a high level of aluminium (Baker & Schofield, 1982). However, the effects of acidification on organisms begin at pH values of about 6 (Psenner & Zapf, 1990).

The number of lakes with pH values below 5.3 and below 6 are also reported in Table 2 with their 95% confidence interval; pH values below 5.0 were also found in 31 out of 806 total analyses (4%).

Factors affecting lake chemistry

Relations between chemical variables were defined using factor analysis on standardized concentrations, excluding pH because it is not truly independent of the other variables, and considering nitrate and ammonium together to take into account the redox processes that can transform them into each other.

Table 3. Varimax rotated factor matrix.

Factor	1	2	3	4
Accounted variance	38.3%	18.9%	10.8%	9.8%
Ca^{2+}	0.91	0.04	0.15	−0.24
Mg^{2+}	0.89	0.10	0.09	0.01
Alkalinity	0.73	−0.02	−0.05	0.14
SO_4^{2-}	0.88	0.13	0.02	0.19
Na^+	−0.07	0.83	0.16	0.12
Reactive Si	0.27	0.75	0.28	−0.41
K^+	0.32	−0.05	0.76	−0.06
Cl^-	−0.08	0.31	0.78	0.19
Inorganic N	−0.07	0.06	0.11	0.93

The results show that the first 4 factors account for 77.8% of the total variance. Their loadings, after Varimax rotation, are showed in Table 3. We consider the first factor, loading on Ca, Mg, sulphate and alkalinity, and the second, loading on Na and silica, as weathering factors, for carbonatic and acidic rocks, respectively. So weathering accounts for more than one half of the total variability. The first two factors are both significantly related to the lithology of the watersheds, as shown by Kendall rank correlation analysis performed between factor scores and a semiquantitative lithological description in which lower values are assigned to acidic rocks and higher values to carbonate rocks (Mosello *et al.*, 1991).

For the third factor, loading on K and chloride, the interpretation is less simple. It is significantly related to the amount of vegetation in the watershed and, inversely, to its surface. It may be an evapotranspiration and ionic exchange factor.

The last factor, directly related to total nitrogen and inversely related to the amount of vegetation in the watershed, represents the nitrogen uptake by the watershed vegetation.

Only the first and fourth factors are related to alkalinity, indicating that the latter is mainly controlled by carbonate weathering and nitrogen uptake.

Level of acidification

The preacidification value of total alkalinity, TAo, was estimated from pristine base cation concentrations, (Ca + Mg)o, on the basis of the empirical relationship (Wright, 1983):

$$TAo = 0.91\ (Ca + Mg)o$$

Sea-salt correction of ionic concentrations is omitted, because of the low chloride level.

Pristine base cation concentrations can be estimated from present values assuming no change in weathering rate (Wright, 1983) or an increase in weathering and cation desorption caused by the deposition of strong acid. In this case it would be buffered by the increase in base cation concentrations, $\Delta(Ca + Mg)$, related to those of strong acid anions, in our case $\Delta(SO_4 + NO_3)$, through the variable F-factor (Brakke *et al.*, 1990):

$$F = \frac{\Delta\,(Ca + Mg)}{\Delta\,(SO_4 + NO_3)} = \sin\left(90^\circ\ \frac{(Ca + Mg)}{S}\right),$$

where S was set to 200–400 $\mu eq\ l^{-1}$ depending on the acid loading. If the base cation concentrations exceed S, the acid input will be completely buffered by the watershed weathering.

The maximum value of F can be estimated by assuming that the initial base cation concentration was zero and that the background sulfate (and nitrate) concentration is about 8–15 $\mu eq\ l^{-1}$ (Brakke *et al.*, 1990) or negligible: in this case F will be smaller than $(Ca + Mg)/(SO_4 + NO_3)$.

Table 4. Loss of alkalinity ($\mu eq\ l^{-1}$) for lakes with base cation concentration below S ($\mu eq\ l^{-1}$), assuming different values of S.

Values of S assumed	200	400	∞
No. of lakes	42	55	125
25th percentile	9	21	38
Median	18	32	56
75th percentile	26	40	73
Average	17	30	89

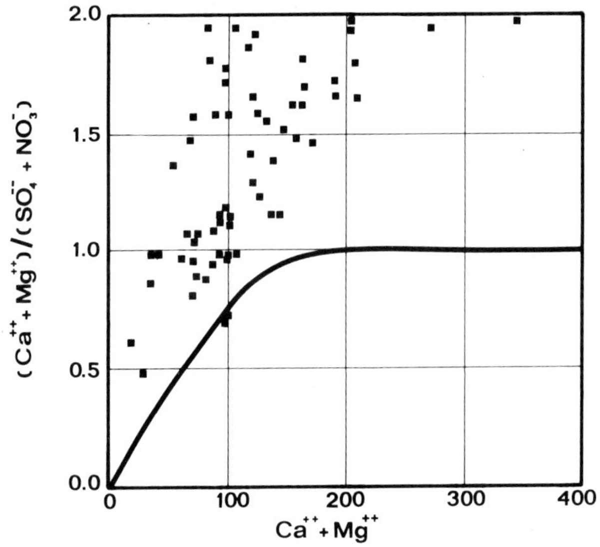

Fig. 3. Ratio of base cation concentration (Ca + Mg) to those of strong acid anion ($SO_4 + NO_3$) plotted versus base cation concentration (Ca + Mg) and compared with calculated F values, assuming $S = 180$. Unit: $\mu eq\ l^{-1}$.

With this set of data, the minimum S value is about 180 (Fig. 3), confirming Brakke's approach.

The acidification levels (losses of alkalinity) calculated assuming $S = 200$ and $S = 400$ are compared with the values obtained by assuming no change in weathering rate (S set to ∞) in Table 4.

Quantification of processes leading to alkalinity production

For lakes lying in watersheds formed of carbonatic rocks, carbonate weathering accounts for nearly all the production of alkalinity, but in lakes with watersheds completely formed of acidic rocks, silicate weathering and nitrate uptake may play a major role.

To find out whether the above-mentioned processes can account for all the alkalinity production, we focused on nineteen lakes in the Maggia Valley, Switzerland, sampled five times during summer 1988. Since the valley is located only 10 km east of the atmospheric deposition sampling station of lake Toggia (Fig. 1), these lakes are suitable for a comparison between atmospheric deposition and lake chemistry. In order to avoid episodic fluctuations, the ionic composition of each lake was obtained by choosing the median concentration of each solute. The Maggia Valley is one of the most acid-sensitive areas in the Alps, so these lakes are not representative of the whole set of lakes studied, lying as they do mainly on granitic bedrock: their alkalinity is below 50 $\mu eq\,l^{-1}$, with a median value of 2 $\mu eq\,l^{-1}$.

Table 5 shows a comparison between lake chemistry and atmospheric deposition, corrected for evapotranspiration by multiplying for the factor precipitation/runoff of 1.4 proposed for this area by Giovanoli *et al.* (1988).

On this basis, we calculated the alkalinity for each lake on the basis of the following assumptions:

- sulphate and sulphide production is of minor importance, as they produce less than 5 $\mu eq\,l^{-1}$ of sulphate in 72% of the lakes;
- the alkalinity produced by in-lake processes is negligible;
- ammonium is completely oxidised to nitrate, producing two equivalents of acidity, or ammonium is in part taken up producing one equivalent: because of an equivalently lower uptake of nitrate, there is no variation in the resulting alkalinity production and the calculation is also valid for all intermediate situations;

Table 5. Comparison between atmospheric deposition and lake chemistry (unit: $\mu eq\,l^{-1}$), silica in $\mu mol\,l^{-1}$.

Variable	Bulk deposition	Deposition corrected	Lake range	Lake median
pH	4.88	–	4.82–6.73	5.86
H^+	13	18	0–15	1
T.A.	−13	−18	−15–49	2
NH_4^+	22	30	1–3	1
Ca^{2+}	20	28	38–92	55
Mg^{2+}	5	7	4–15	14
Na^+	5	7	8–24	14
K^+	2	3	3–13	6
SO_4^{2-}	33	46	31–60	50
NO_3^-	19	27	14–39	23
Cl^-	7	10	5–8	5
R. Si	0	0	16–52	28

- nitrate uptake from vegetation produces one equivalent of alkalinity;
- silica losses through sedimentation of diatoms are negligible, because oligotrophy and high runoff limit primary production, and pulses of low silica concentration caused by chrysophycean uptake were avoided during data screening;
- silicate minerals are weathered to gibbsite, producing alkalinity: alkalinity production can be evaluated from 0.6 equivalent (plus 0.08 equivalent of Ca) as calculated for rocks in the same area by Giovanoli *et al.* (1988) to 1 equivalent, as proposed by Zobrist *et al.* (1987). The first value was chosen because it provides a more prudent estimate of alkalinity production;
- the remaining increase in Ca is due to the weathering of calcite, which is present in small quantities, but highly soluble, and produces one equivalent of alkalinity;
- the apparent chloride sink is due to slight contamination of snow samples.

In each lake, alkalinity is then calculated as:

$$\begin{aligned} Alk = & -1.4\,[H^+]_{deposition} + \\ & + (1.4\,[NO_3^-]_{deposition} - [NO_3^-]_{lake}) - \\ & - (1.4\,[NH_4^+]_{deposition} - [NH_4^+]_{lake}) + \\ & + 0.6\,[Si]_{lake} + ([Ca^{++}]_{lake} - 0.08\,[Si]_{lake} - \\ & - 1.4\,[Ca^{++}]_{deposition}) \end{aligned}$$

The differences between observed and calculated alkalinities (Fig. 4) represent the alkalinity produced by other processes, such as ionic exchange, in-lake production, Mg-carbonate weathering and aluminium hydroxide dissolution. They range between −1 and 21 $\mu eq\,l^{-1}$, with 90% of the values below 12 $\mu eq\,l^{-1}$ (17% of the total alkalinity production). However, in two out of the 19 lakes, Lower and Upper Laghetto, unexplained alkalinity production reaches 21 and 27% respectively. They have longer renewal time (about one year) than typical values for Alpine lakes in this area (a few weeks): losses in silica caused by diatom sedimentation may therefore play an important role, causing an underestimate of silicate mineral

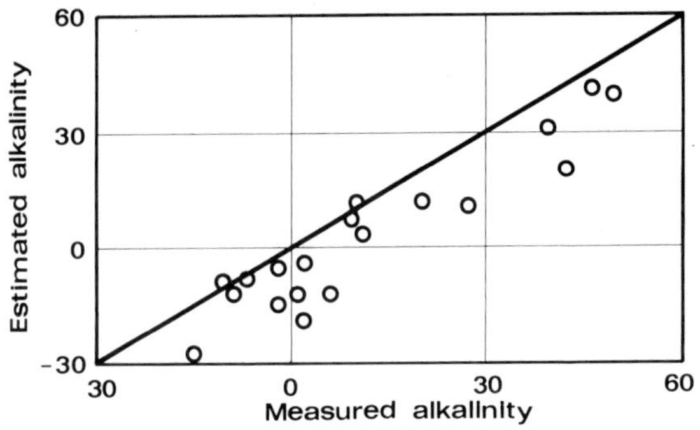

Fig. 4. Differences between measured alkalinity and the value calculated considering only watershed processes. Unit: μeq l^{-1}

weathering that may account for half of the unexplained alkalinity production.

The oxidation of the ammonium deriving from atmospheric deposition is an important source of acidity, and can contribute up to 62% of the acidifying capacity of deposition itself.

Conclusions

In spite of the episodic deposition of Saharian dust, Alpine lakes in the Southern Alps receive an atmospheric input of pollutants, mainly sulphate, nitrate and ammonium, which cause acidification processes. Along with acid loading, ammonium oxidation is an important source of lake acidity.

Calcite and silicate weathering and nitrogen uptake from vegetation are the main sources of alkalinity buffering the atmospheric acid loading. In carbonate watersheds, weathering is the main source of alkalinity, but calcite weathering is also present in acidic watersheds and watershed alkalinity production accounts typically for 89% or more of the total production. However, in some lakes other processes may contribute up to 27% of the alkalinity.

The alkalinity producted is not enough to avoid acidification processes, as 11 lakes show summer pH of below 5.3 and 18% of the pH values measured were lower than 6.0. In front of a deposition potential acidity of 57 μeq l^{-1}, summer acidification levels of below 60 μeq l^{-1} were calculated for acid-sensitive lakes using a titration model with variable F-factor with median values ranging from 18 to 32 μeq l^{-1}, depending on parameter setting.

References

Barbieri, A. & G. Righetti, 1987. Chimica delle deposizioni atmosferiche nel Cantone Ticino ed effetti sulle acque dei laghi alpini d'alta quota. Documenta Ist ital. Idrobiol. 14: 19–34.

Baker, J. P. & C. L. Schofield, 1982. Aluminium toxicity to fish in acidic waters. Wat. Air Soil Pollut. 18: 289–309.

Brakke, D. F., A. Henriksen & S. A. Norton, 1990. A variable F-factor to explain changes in base cation concentrations as a function of strong acid deposition. Verh. int. Ver. Limnol. 24: 146–149.

Carollo, A., F. Contardi, V. Libera & A. Rolla, 1985. Hydroclimatic cartography of the Lake Maggiore drainage basin. Mem. Ist. ital. Idrobiol. 42: 1–4.

Giovanoli, R., J. L. Schnoor, L. Sigg, W. Stumm & J. Zobrist, 1988. Chemical weathering of crystalline rocks in the catchment area of acidic Ticino lakes, Switzerland. Clays and Clay Minerals 36: 521–529.

Goldstein, R. A. & S. A. Gherini (eds), 1984. The integrated lake-watershed acidification study. Report EPRI EA-3221.

Gruppo di Studio, 1987. Deposizioni atmosferiche nel Nord Italia. Rapporto finale anni 1983–84. Quaderni Ingegneria Ambientale 6: 5–63.

Henriksen, A., 1980. Acidification of freshwater. A large scale titration. In D. Drablos and Tollan, A. (eds), Proc. Int. Conf. on Ecol. Impact of Acid Precipitation. As, Norway: 68–74.

Mosello, R., 1984. Hydrochemistry of high altitude alpine lakes. Schweiz. Z. Hydrol. 46: 86–99.

Mosello, R., A. Marchetto, G. A. Tartari & L. Guzzi. 1988. Acidificazione e acidificabilità delle acque lacustri italiane. Documenta Ist. ital. Idrobiol. 15: 1–80.

Mosello, R., A. Marchetto, G. A. Tartari, M. Bovio & P. Castello, 1991. Chemistry of Alpine lakes in Aosta Valley (N. Italy) in relation to watershed characteristics and acid deposition. Ambio 20: 7–12.

Psenner, R., K. Arzet, A. Brugger, J. Franzoi, F. Heisberger, W. Honsig-Erlenburg, F. Horner, U. Nickus, P. Pfister, P. Schaber & F. Zapf, 1988. Versauerung von Hockgebirgsseen in Kristallinen Einzugsgebieten Tirols und Kaerntens. Bundesministerium fuer Land- und Forstwirtschaft, Wien, 335 pp.

Psenner, R. & F. Zapf, 1990. High mountain lakes in the Alps: peculiarity and biology. In: M. Johannessen, R. Mosello and H. Barth (eds), Acidification processes in remote mountain lakes. Guyot, Brussels: 22–38.

Sullivan, T. J., D. F. Charles, J. P. Smol, B. F. Cumming, A. R. Selle, D. R. Thomas, J. A. Bernert & S. S. Dixit, 1990. Quantification of changes in lakewater chemistry in response to acidic deposition. Nature 345: 54–58.

Wright, R. F., 1983. Predicting acidification of North American lakes. EPA, Corvallis, Oregan: 165 pp.

Zobrist, J., Sigg, L., Schnoor, J. L. & W. Stumm, 1987. Buffering mechanism in acidified Alpine lakes. In: Barth, H. (ed.), Reversibility of acidification. Elsevier, London: 95–103.

Hydrobiologia **274**: 83–90, 1994.
J. Fott (ed.), Limnology of Mountain Lakes.

Reconstruction of pH by chrysophycean scales in some lakes of the Southern Alps

Aldo Marchetto & Andrea Lami
CNR Istituto Italiano di Idrobiologia, I-28048 Verbania Pallanza, Italy

Key words: acidification, *Mallomonas*, paleolimnology, scales, sediment

Abstract

Chrysophycean scales were examined in surface sediments collected from 22 high mountain lakes on the southern slope of the Central Alps, some in Italy and some in Switzerland. The study area receives slightly acidic precipitation and summer lake pH ranges between 5.2 and 8.0.

In each lake chrysophycean scale assemblage was dominated by one or two species and its composition was related to lakewater pH.

Five short cores were examined in low-alkalinity lakes and evidence of recent lake acidification was found.

Introduction

In the last decade the aim of acidification research has moved from identifying areas receiving acid deposition to understanding acidification processes in order to predict how ecosystems respond to acidification. In this context, paleolimnological techniques have been regarded as a tool for investigating changes in lake ecosystems in order to find out if these changes are coupled with acid deposition.

Several variables can be analysed in sediment cores; among them, diatom remains have been widely used to reconstruct historical pH changes (for a review see Charles *et al.*, 1989). Besides diatoms, chrysophycean scale assemblage has been used to infer past lake pH in North America (Smol, 1986) and Europe (Hartmann & Steinberg, 1986). This technique may have some advantages over the use of diatoms in describing early stages of acidification, because most of the scale-bearing chrysophytes are euplanktonic, while many diatom species are benthonic and so respond with some delay to acidification (Charles *et al.*, 1989). On the other hand, problems may arise from the small number of chrysophyte species generally found in lakes.

The aim of this paper is to establish whether chrysophycean scale assemblages can be used to detect early acidification in high mountain lakes in the Alps, where these algae have up to now been rarely reported (e.g. Nebaeus, 1984; Giussani *et al.*, 1986).

In this paper we develop region-specific calibration equations and discuss the suitability of different methods to infer past pH by chrysophycean remains in high mountain lakes. The most significant transfer function is then used to infer pH histories from cores collected in five lakes.

Study lakes

Sediment samples were collected in 27 high mountain lakes in the Southern Alps, some in

Asmund & Cronberg. In alkaline lakes (pH 7.9–8.0) the main species were *M. acaroides* var. *acaroides* Perty em. Ivanoff (T) and *M. alpina* Pascher & Ruttner em. Asmund & Kristiansen (BI). In acid lakes the most common scales belonged to small forms, namely *M. alveolata* Dürrschmidt (PS, Sfo) and *M. actinoloma* var. *maramuresensis* Péterfi & Momeu (C, La, MS, VI, VS, Z), apart from one lake (Le) in which there were as many scales of *Chrysosphaerella brevispina* Korschikov em. Harris & Bradley as those of *M. multisetigera* Dürrschmidt. One lake (MI) was located near a small goat-fold and was dominated by *Synura echinulata* Korschikov.

In core analysis it was found that chrysophycean scales were well preserved in the 2–10 top most slices, while they were not present in deeper sediments. Therefore, for pH reconstruction along sediment cores we chose five lakes (Le, P, MS, PS and VI) whose sediments were dominated by common species and had enough scales for four or more slices. In four of them, the abundance of *M. crassisquama* scales was higher in deeper slices than in topmost sediments (Fig. 2–5), while no change was found in Lake Panelatte (Fig. 6), which shows higher pH (7.21) and alkalinity (0.11 meq l^{-1}).

Reconstructed pH histories

Transfer equations were developed using several methods on a reduced subset generated by excluding species with a frequency of occurrence lower than 5% or occurring in less than 3 lakes. Five lakes dominated by these species were also excluded.

The methods used were:

(a) multiple regression using taxa (Charles, 1985);
(b) weighted averaging (ter Braak & Barendregt, 1986);
(c) weighted averaging, using an extra regression (ter Braak & van Dam, 1989) to avoid the centroid tendency of that method;
(d) regression between pH and the first axis of canonical correspondence analysis (ter Braak, 1987);

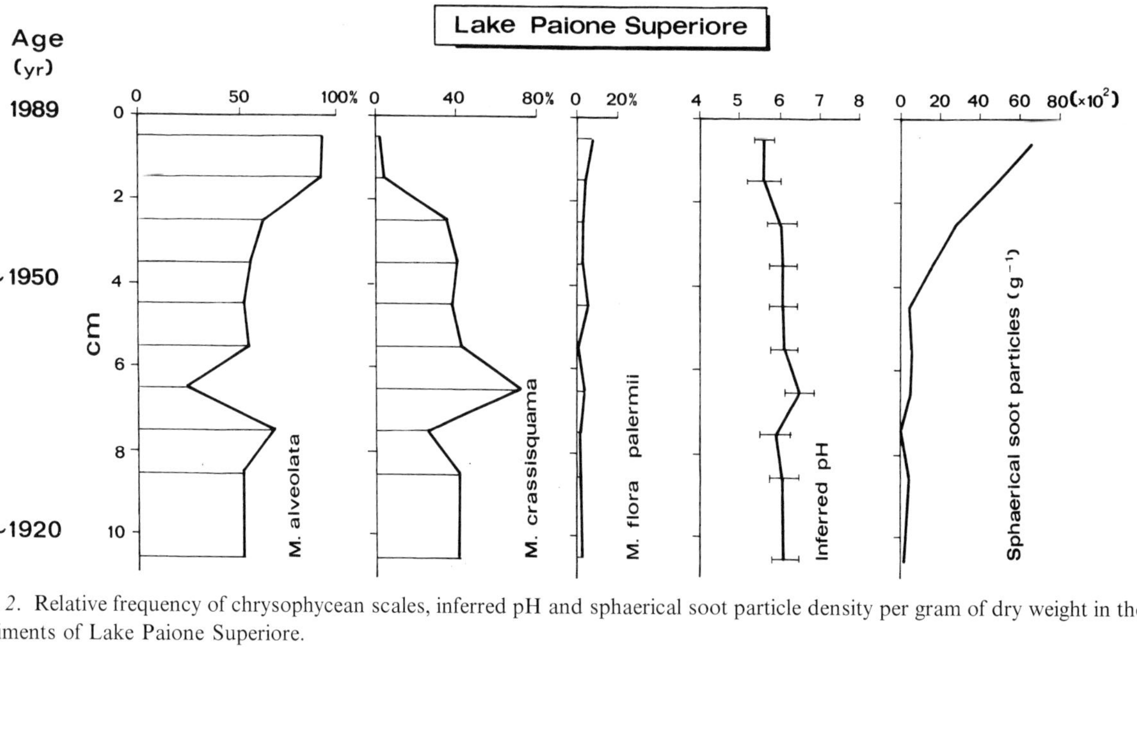

Fig. 2. Relative frequency of chrysophycean scales, inferred pH and sphaerical soot particle density per gram of dry weight in the sediments of Lake Paione Superiore.

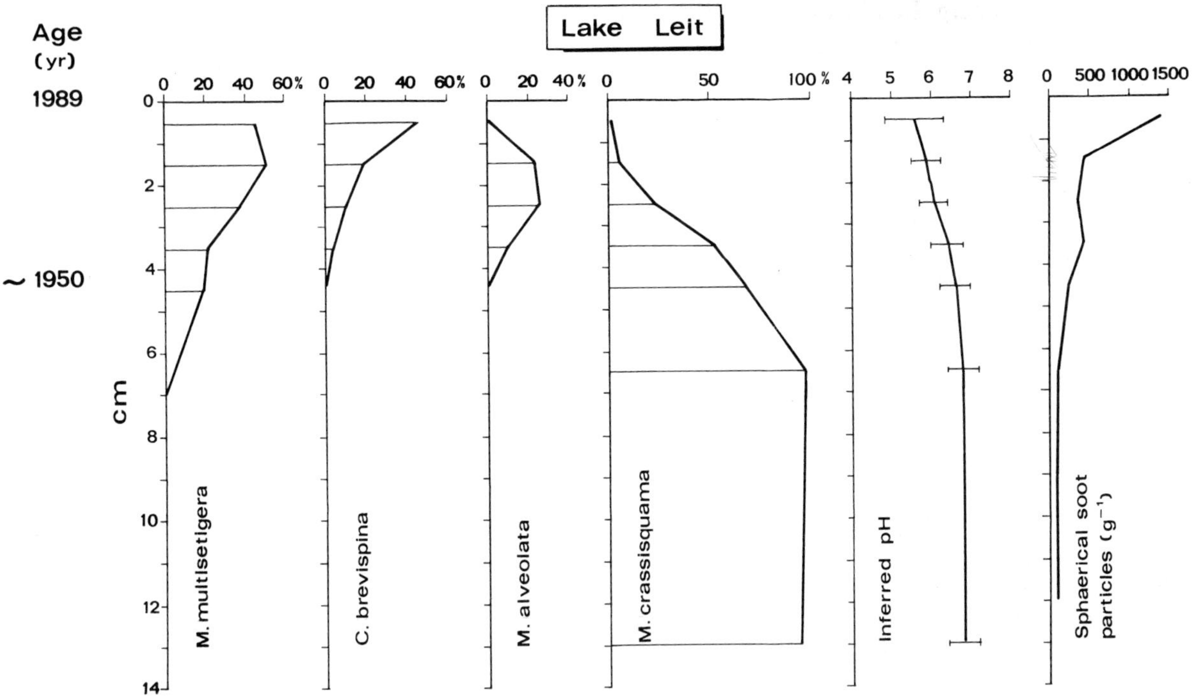

Fig. 3. Relative frequency of chrysophycean scales, inferred pH and sphaerical soot particle density per gram of dry weight in the sediment of Lake Leit.

(e) multiple regression using preference group (Charles & Smol, 1988).

Inferred pH values derived from each method were compared with contemporary pH of reduced subset (Table 3), with similar results: weighted averaging with extra regression performed best (Fig. 7) and was used for core analyses. Inferred pH values (Table 4) are lower in the upper section of each core. The standard error of the pH esti-

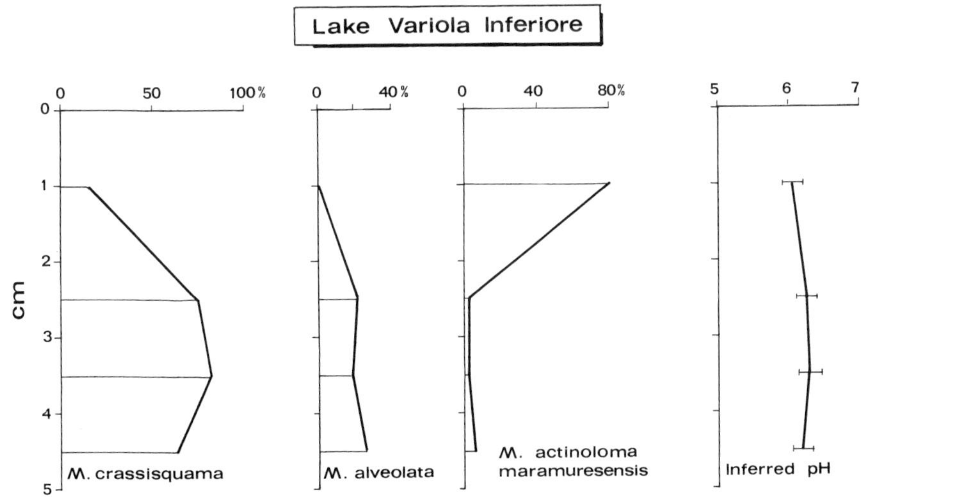

Fig. 4. Relative frequency of chrysophycean scales and inferred pH in the sediments of Lake Variola Inferiore.

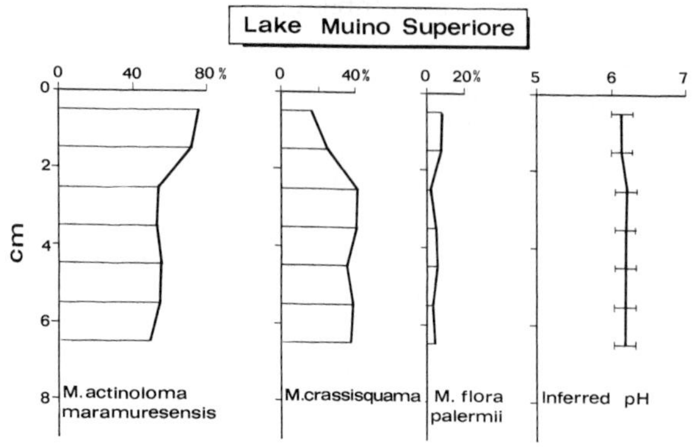

Fig. 5. Relative frequency of chrysophycean scales and inferred pH in the sediments of Lake Muino Superiore.

Table 3. Comparison of methods for pH-reconstruction.

Method	Intra-set standard error
a	0.42
b	0.46
c	0.41
d	0.42
e	0.45

mate was evaluated using a bootstrap procedure (Birks *et al.*, 1990) with 1000 cycles: in each cycle a training subset of the same size as the original data set was randomly selected and the samples which were not selected formed a test set. Surface sample standard errors were estimated as the standard errors of prediction across all test sets. Fossil sample standard errors were evaluated as the sum of squares of the standard deviation of the prediction across all test sets and a constant component representing variations in taxon abundances. This component was estimated for each test sample by the root mean square of the difference between observed pH and bootstrap pH across the training sets that do not contain this sample.

The point of change of the pH profiles was evaluated using a maximum likelihood procedure (Esterby & El-Shaarawi, 1981). Mean pH and its standard error were then evaluated for the two sections of each core lying above and below the point of change (Table 4). The differences between the upper and lower sections were significant ($P<0.01$; Student-t test) for Lakes Paione Superiore (PS) and Leit (Le), only.

Finally, the overall significance of the pH decrease in the five lakes was evaluated using the

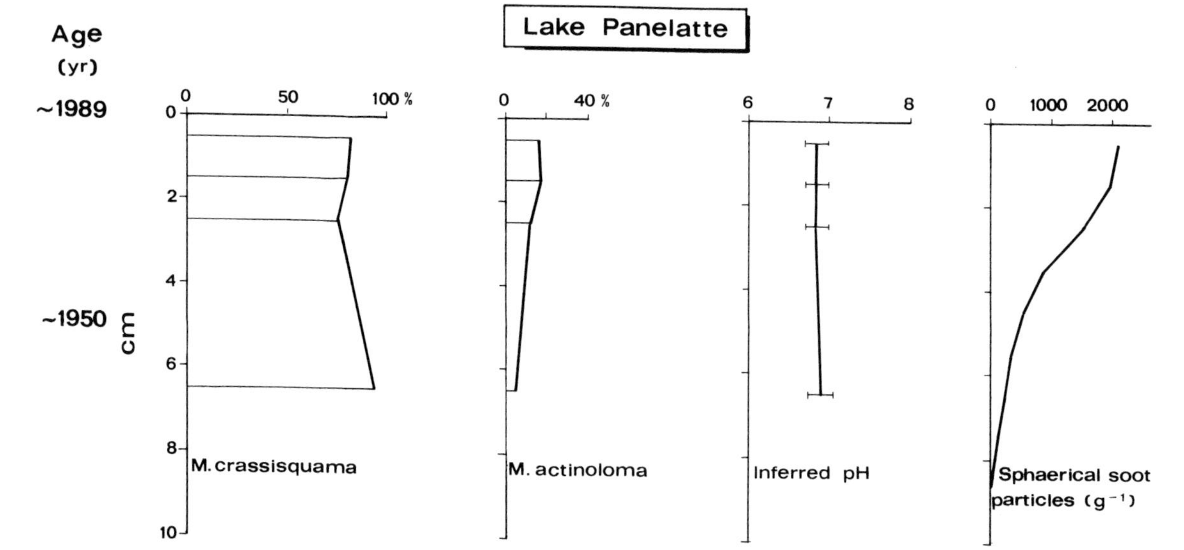

Fig. 6. Relative frequency of chrysophycean scales, inferred pH and sphaerical soot particles density per gram of dry weight in the sediments of Lake Panelatte.

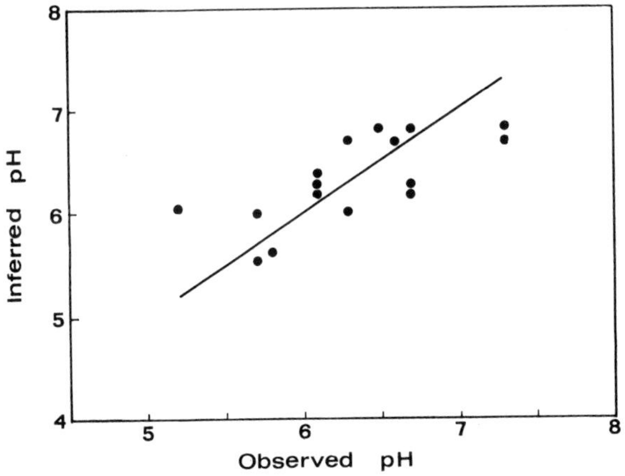

Fig. 7. Plot of chrysophycean inferred pH versus measured summer pH.

technique proposed by Fisher (1954): the result was a highly significant pH decrease ($P<0.01$).

Discussion

Diatoms have been successfully used to infer past lakewater pH, but in Alpine lakes scaled chrysophyceans seem to perform better because they are all planktonic and therefore not influenced by the buffering effect of bottom sediments, and they develop in spring when lake acidity is higher.

Table 4. Measured and inferred pH with standard error.

Lake	PS	L	MS	VI	P
Measured pH (Autumn)	5.72	6.31	6.08	6.65	7.26
Point of change cm	2	4	1	2	–
Upper section:					
No. of slices	2	4	1	1	–
Mean pH	5.66	5.99	6.29	6.08	–
Standard error	0.21	0.25	0.33	0.79	–
Lower section:					
No. of slices	8	3	3	3	4
Mean pH	6.14	6.72	6.40	6.52	6.70
Standard error	0.13	0.12	0.21	0.21	0.18
pH decrease	0.48*	0.73*	0.11	0.34	–

* Significant at 0.01 level.

However, the small number of species could be a bar to devising transfer functions. Moreover, a lack of knowledge of their seasonal development and the virtual absence of pH measurement under the ice and at the snowmelt forced us to use late summer pH in refining calibration equations. Finally, low lake sedimentation rates can result in differences between contemporary chrysophycean communities and scale assemblage in the topmost slices of sediment cores. In spite of these problems, the standard error was moderate (less than 0.4).

Four poorly-buffered lakes showed evidence of recent acidification which, however, was statistically significant in two lakes only. The fact that the timing of the inferred acidification corresponds to a sharp increase in industrially produced carbonaceous particles points to acid deposition as the probable cause. Alternative hypotheses could be changes in land use in the catchments and post glacial natural changes. The lakes selected for coring are all above the timberline, so acidification caused by forest development can be excluded. They are also far from human disturbances, apart from goat grazing. Humification of the lake catchment is unlikely for these clearwater lakes. Fish stocking could modify planktonic communities and affect pH estimation, but the inferred acidification is similar in poorly-buffered lakes with and without fishes.

The effect of post glacial natural changes needs further investigation in Alpine sectors receiving less polluted deposition, but it seems inconsistent with the simultaneous increase in all the lakes of both acidity and carbonaceous particles.

Conclusions

Chrysophycean scale assemblages can be used to develop transfer functions to infer the past acidity of high mountain lakes. Inferred pH trends in four poorly-buffered lakes on the southern slope of the Central Alps suggest recent lake acidification. This is consistent with the composition of atmospheric deposition in this area, the acidity and ammonium concentration of which are of sufficient magnitude to affect more sensitive freshwaters.

Acknowledgements

The study on carbonaceous particles was supported by the CNR-ENEL project-Interactions of Energy Systems with Human Health and Environment-Rome, Italy. Absolute core dating using ^{210}Pb was provided by Dr L. Guzzi (ENEL-CRTN, Milan).

References

Birks, H. J. B., J. M. Line, S. Juggins, A. C. Stevenson & C. J. F. ter Braak, 1990. Diatoms and pH reconstruction. Phil. Trans. r. Soc., Lond. 327: 263–278.

Charles, D. F. & J. P. Smol, 1988. New methods for using diatoms and chrysophytes to infer past pH of low-alkalinity lakes. Limnol. Oceanogr. 33: 1451–1462.

Charles, D. F., R. W. Battarbee, I. Renberg, H. van Dam & J. P. Smol, 1989. Paleoecological analysis of lake acidification trends in North America and Europe using diatoms and chrysophytes. In S. A. Norton, S. E. Lindberg & S. L. Page (eds), Acidic precipitation. 4. Soils, aquatic processes and lake acidification. Springler-Verlag, New York: 207–275.

Esterby, S. R. & A. H. el-Shaarawi, 1981. Likelihood inference about the point of change in a regression regime. J. Hydrol. 53: 17–30.

Fisher, R. A., 1954. Statistical methods for research workers. Olyver and Boyd, Edimburgh, 138 pp.

Giussani, G., R. de Bernardi, R. Mosello, I. Origgi & T. Ruffoni, 1986. Indagine limnologica sui laghi alpini d'alta quota. Documenta Ist. ital. Idrobiol. 9: 1–415.

Hartmann, H. & C. Steinberg, 1986. Mallomonadacean (Chrysophyceae) scales. Early biotic paleoindicators of lake acidification. Hydrobiologia 143/Dev. Hydrobiol. 37: 87–92.

Marchetto, A., A. Barbieri, R. Mosello & G. A. Tartari, 1994. Acidification and weathering processes in high mountain lakes in Southern Alps. Hydrobiologia 274/Dev. Hydrobiol. 93: 75–81.

Nebaeus, M., 1984. Algal water-blooms under ice cover. Verh. int. Ver. Limnol. 22: 719–724.

Rendberg, I. & M. Wik, 1984. Dating recent lake sediments by sot particle counting. Verh. int. Ver. Limnol. 22: 712–718.

Smol, J. P., 1986. Crysophycean microfossils as indicators of lakewater pH. In J. P. Smol, R. W. Battarbee, R. B. Davis & J. Meriläinen (eds), Diatoms and lake acidity. Dev. Hydrobiol. 29. Dr W. Junk Publishers, Dodrecht: 257–287.

ter Braak, C. J. F., 1987. Calibration. In R. H. G. Jongman, C. J. F. ter Braak & O. F. R. van Tongeren (eds), Data analysis in community and landscape ecology. Pudoc, Wageningen, The Netherlands: 78–90.

ter Braak, C. J. F. & L. G. Barendregt, 1986. Weighted averaging of species indicator value: its efficiency in environmental calibration. Math. Biosci. 78: 57–72.

ter Braak, C. J. F. & H. van Dam, 1989. Inferring pH from diatoms: a comparison of old and new calibration methods. Hydrobiologia 178: 209–223.

Hydrobiologia **274**: 91–100, 1994.
J. Fott (ed.), Limnology of Mountain Lakes.

A survey on water chemistry and plankton in high mountain lakes in northern Swedish Lapland

Arnold Nauwerck
Limnological Institute, Austrian Academy of Sciences, A-5310 Mondsee, Austria

Key words: acidification, mountain lakes, Lapland

Abstract

A helicopter survey was carried out on 56 water bodies in the Abisko mountains, Swedish Lapland, in August 1981. Water chemistry was found to be highly correlated with bedrock quality in the drainage area of the lakes. Low pH values (down to 5.1) appeared in the neighbourhood of sulphuric iron-ores. Natrium and chloride concentrations showed large scale patterns which can be explained by orographic rainfall. Biologically, northern high mountain conditions are reflected in species composition rather than in biomass or possibly in diversity. Small chrysomonades and dinoflagellates, as well as *Keratella hiemalis* and *Cyclops scutifer* characterize the most 'arctic' waters. A comparison with data from earlier investigations did not confirm expected signs of acidification.

Introduction

High mountain lakes are supposed to be particularly sensitive to environmental changes. Because of small drainage areas, with limited chemical and biological erosion and extreme climatic conditions, they develop simple and labile ecosystems which react promptly to environmental stress and which makes them suited to function as early warning systems. It was therefore of interest to learn more about the characteristics of northern Scandinavian mountain lakes and, as there are indications of a progressing acidification also in the north of Scandinavia, to get a preliminary answer to the questions of what the chemical and biological state of high mountain lakes of the northern Scandes is and if there are any signs of acidification?

In August 1981 a helicopter survey was carried out by the county administration of Norrbotten, supported by the Environmental Agency in Stockholm, and some 60 lakes in the Abisko area (Fig. 1) were sampled. The results have been published as a stencilled report (Nauwerck, 1983). With respect to the continued discussion, it seems justified to make them available to a broader public, as a background for later investigations and as comparative material.

Earlier information on water chemistry from the area stems from Ohle (1940), Ekman *et al.* (1950) Rodhe (1963), Rodhe *et al.* (1966), Holmgren (1968), Eriksson & Persson (1971), Jansson (1972–1978), Nauwerck & Ramberg (1979), Nauwerck (1981). Phytoplankton has been studied by Skuja (1964), Nauwerck (1966), Holmgren (1968), Nauwerck & Persson (1971), Nauwerck & Ramberg (1979) and others. Zooplankton investigations have been carried out by Ekman (1904), Pejler (1957), Eriksson & Persson (1971). A general description of the Abisko lakes and their condition is given by Ekman (1957). Particularly Lake Latnjajaure has been the object

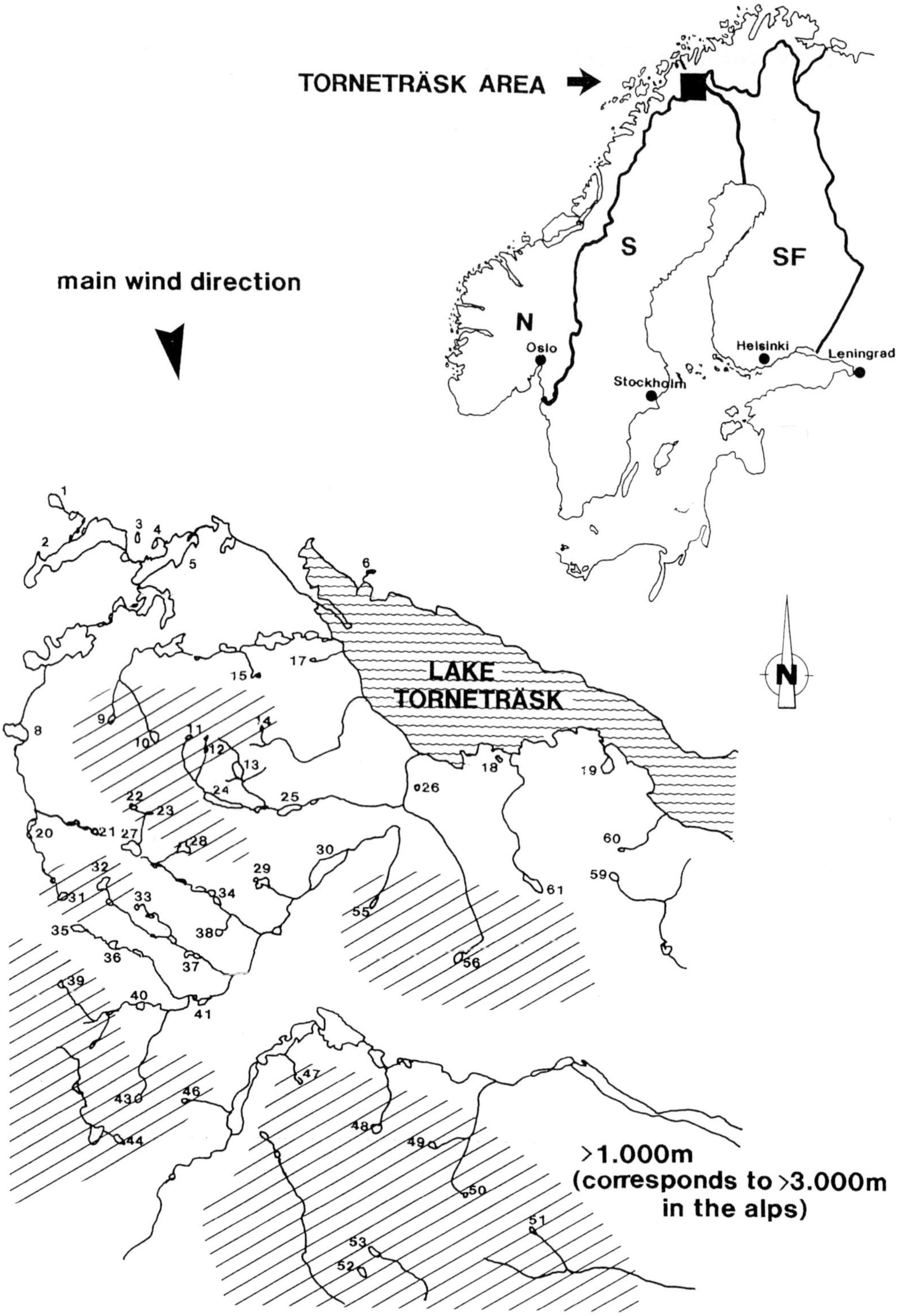

Fig. 1. The Abisko region south of Lake Torneträsk in Swedish Lapland. Hatched area: more than 1000 m above sea level.

of intense studies and much general information can be gained from it (Nauwerck 1968, 1978, 1980). Comparable data exist from southern Sweden and from high mountain lakes elsewhere in the world, e.g. Andersson (1971), Caplanqc (1972), Larson (1973), Almer *et al.* (1974), Raddum *et al.* (1980), Honsig-Ehrenburg & Psenner (1986), Mosello (1986), Psenner *et al.* (1988).

Material and methods

It is an *a priori* of the survey, that the geology of the region is very variable. At the southern edge of the investigated area there are amphibolite stocks; a large zone of mica schist stretches from Abisko towards the west; granite dominates in the valley between Lake Torneträsk and the Norwegian border and hard schists are common in the drainage area of the Abisko River. Exposures of limestone are found at the northern edge of the area and around the Abisko mica schists. The lake's names and numbers are given in Table 1. Their sea level is between 348 m and 1344 m. Most of the lakes are situated above the tree-line, which in the Abisko region is between 600 and 700 m above sea level. It may be mentioned that 1000 m climatically and biologically, roughly correspond to 3000 m in the Alps.

The survey took place during 19th and 20th August 1981. Because of technical reasons (helicopter time!) samples were collected from the surface only. This sets limitations to the data, as it is known that in high mountain lakes much of the pelagical life is bound to the bottom water layers (Nauwerck, 1966). Secchi depth and surface temperatures were measured in the field. Phytoplankton samples were fixed with Lugols solution for later counting with the help of an inverted microscope. Zooplankton was netted quantitatively with a 100 μm plankton net and was preserved with formaline. For chlorophyll analysis, 3–5 liters of water were taken to the air base and were filtered at the end of every flight, and the dried filters were brought back to the laboratory in freezer boxes for further treatment.

Water chemistry was analyzed by standard methods (Ahl 1972). Most analyses were carried out by the Environmental Agency's limnological laboratory in Uppsala. Plankton and chlorophyll as well as parts of the water chemistry were analyzed by the county administration (*i.e.* A. Nauwerck, L. Lindquist & H. Groth).

Primary data are taken from Nauwerck (1983). For this presentation they are grouped geographically or according to different heights above sea level. Comparisons are made with earlier data so far as such are available. Direct comparisons with other high mountain areas are difficult because of basic differences in natural geography and of very varying approaches. However, general principles can be compared. The closest comparison being the work of Psenner *et al.* (1988) from high alpine lakes.

Table 1. Lake numbers and sea level.

1. Jappnjajaure (484 m)	30. Abiskojaure (487 m)
2. Pajeb Njuorajaure (435 m)	31. Valfopadajaure (1016 m)
3. Kanisjaure (456 m)	32. Kärpeljaure (1005 m)
4. Felisjaure (455 m)	33. Tjåtnjaranjaure (970 m)
5. Vuolep Njuorajaure (426 m)	34. Äparasjaure (766 m)
6. Tjetnjajaure (371 m)	35. Tåresjaure (946 m)
8. Katterjaure (697 m)	36. Valfojaure (835 m)
9. Ekosjön (1110 m)	37. Suonjetjaure (853 m)
10. Rissajaure (815 m)	38. Tjamojaure (860 m)
11. Kuoblatjåkkojaure (1300 m)	39. Ruovsujaure (1010 m)
12. Måndalssjön (1173 m)	40. Snarapjaure (650 m)
13. Latnjajaure (978 m)	41. Kamasjaure (609 m)
14. Ekmanjaure (1235 m)	43. Palep Snarapjaure (1040 m)
15. Raikejaure (902 m)	44. Kartejaure (1228 m)
17. '547 m' (547 m)	46. Tjålmeriepejaure (1114 m)
18. Pajeb Tjautjanjarkajaure(382 m)	47. Njuikusjaure (1092 m)
19. Vuoskojaure (348 m)	48. Piegganjaure (1257 m)
20. Tassakenpadajaure (784 m)	49. Skadnjajaure (1304 m)
21. 'Kärpel' (872 m)	50. Vierrojaure (1298 m)
22. 'Håikanjaure' (1015 m)	51. Rassepautajaure (1245 m)
23. '1055 m' (1055 m)	52. Vassatjårrojaure (1344 m)
24. Övre Kårsavaggesjön (696 m)	53. Vasapadajaure (1182 m)
25. Nedre Kårsavaggesjön (670 m)	55. Tjamuhasjaure (1045 m)
26. Nedre Laksjön (409 m)	56. Nissonjaure (1140 m)
27. Håjkanjaure (847 m)	59. Suorojaure (933 m)
28. '1050 m' (1050 m)	60. '865 m' (865 m)
29. Patjujaure (802 m)	61. Tjuonajaure (950 m)

Results

Surface temperatures of the lakes were between 3.2 °C and 14.1 °C. Temperature is dependant on sea level but also on geographical position, light incidence, water colour, throughflow and other factors.

Secchi depth was between 0.7 m in very turbid lake 28, and 16.5 in large lakes 1 and 3 in the northern limestone area. In several lakes Secchi depth was more than bottom depth and could not be measured or had to be approximated. Most lakes had clear or very clear water. Only lakes 18, 19, 26, 60 and 61 in the birch forest region were brownish because of drainage from nearby mires. Usually they also are rich in organic material and show high permanganate consumption.

Conductivity differs between about 140 μS cm^{-1} in little limestone embedded lake 17 and about 6 μS cm^{-1} in the highest situated lake 52. There are slight differences between the analysis results of the county laboratory and the laboratory of the Environmental Agency (the former are generally slightly higher than the latter). In the top lakes 44 and 48–55, conductivity is not more than in rain water.

The top lakes also show low pH values, but not lower than 5.45. Lowest pH value, 5.1 is from lake 12. This lake is characterized by an influence of sulphurous ores. Alkalinity is not measurable in lakes 9, 11, 50, 52, 53. They all have pH values between 5.1 and 5.6, except lake 52 where pH is >6. Values higher than 8 are to be found in high conductivity limestone lakes 17 and 26. More than 60% of all lakes had pH values between 7 and 8, less than 10% had pH values below 6.

Nutrient contents were low or very low. Total phosphorus was between 4 and 12 μg l^{-1}, phosphate-phosphorus in many cases not beeing measurable. The occurrence of apatite in the surroundings of some lakes may be responsible for increased phosphorus values in some cases. (Lake 59 had been 'fertilized' with phosphoric acid by sport fishermen with the intention to improve fish production!) Also nitrate values were low or very low. Phytoplankton productivity can be assumed as a reason for nitrate consumption.

Phytoplankton biomass, expressed as chlorophyll concentration, showed surprisingly little variation while volume calculated fresh weights were more variable. Lowest values of both were found in very clear lake 10 and lake 41 (chl$_a$ 0,15 and 0,18 μg l^{-1}, fresh weight 28,2 and 27,9 μg l^{-1}), highest values were found in top lake 55, in birch forest lake 26 and in mountain lake 38 (chl$_a$ 1,88; 1,37; and 1,26 μg l^{-1}, fresh weight 739,4; 841,9; and 406 μg l^{-1}). Irregularities in chlorophyll/biomass relationships are obviously caused by differences in phytoplankton quality. It is surprising that plankton biomass shows little correlation with the sea level or nutrient level of the lakes. Gradients are instead to be found in plankton composition.

Tables 2a and b show the frequency of the 10 most common species, phytoplankton and zooplankton, in all lakes. Phytoplankton algae *Rhodomonas*, *Spiniferomonas* and *Pseudokephyrion* were found in more than 60% of all lakes but, of course, may be present but rare in others. In zooplankton, the copepode *Cyclops scutifer* was most common, followed by the rotifer *Kellicottia*. The results can be generalized in the following points:

Table 2. Relative frequency of 10 most common phytoplankton and zooplankton species and average number of species in different height levels.

(a) Phytoplankton		(b) Zooplankton	
Rhodomonas minuta	69%	*Cyclops scutifer*	85%
Spiniferomonas sp.	65%	*Kellicottia longispina*	61%
Pseudokephyrion boreale	63%	*Polyarthra vulgaris*	59%
Oocystis submarina v. var.	56%	*Keratella hiemalis*	54%
Chrysolykos Skujai	52%	*Keratella cochlearis*	52%
Pseudokephyrion Entzii	46%	*Conochilus unicornis*	46%
Cyclotella ocellata	46%	*Eubosmina longispina*	44%
Gymnodinium uberrimum	41%	*Daphnia longispina*	35%
Dinobryon acuminatum	41%	*Eudiaptomus graciloides*	35%
Gymnodinium helveticum	33%	*Holopedium gibberum*	26%

(c) average number of species

Altitude	Phytoplankton	Zooplankton
> 1200 m	18.4	7.3
1000–1200 m	22.4	4.9
800–600 m	35.0	8.1
600–400 m	31.4	7.1
< 400	47.3	9.1

Most water quality parameter values increase downhill

Temperature, water colour, permanganate consumption, conductivity, pH value increase on the way from the mountains to the valleys (Figs 2–4). As to the nutrients (Fig. 5), there is little change with sea level in phosphorus, but a decrease in nitrate can be observed from higher to lower lakes.

Relative importance of SO_4^- and Cl^- increases with altitude

As shown by Fig. 3, the ionic composition changes with height. While the composition of the cations remains rather stable, there is a significant relative increase of sulphate with increasing sea level and, in the uppermost lakes even of chloride. In absolute terms the top lakes still have the lowest SO_4 content, 0.032 meq l^{-1}, compared to 0.097 in the lakes below 600 m. However, higher values are to be found in lakes in between.

Bedrock quality and vegetation cause aberrations

As pointed out, the occurrence of sulphurous ores in the immediate vicinity or within the drainage area of some lakes can influence water chemistry as to pH value (lakes 11 and 12, Fig. 4) and also to sulfate (lakes 10, 15, 17). Furthermore, phosphorous content can be influenced by apatite. A limestone background is generally reflected by high conductivity, alkalinity and pH values in the lakes (lakes 17, 26).

Water chemistry also is affected by the vegetation within the drainage area. Lakes surrounded by mires have a high content of humic substances (lakes 18, 19, 60, 61) which means a high total

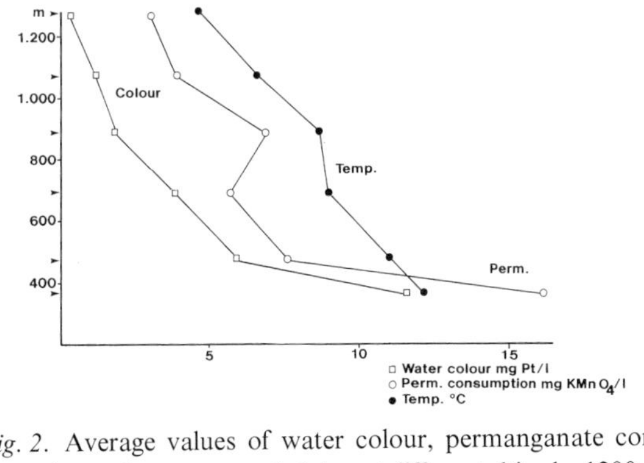

Fig. 2. Average values of water colour, permanganate consumption and temperature in lakes at different altitude: 1200 m above sea level, 1000–1200 m, 800–1000 m, 600–800 m, 400–600 m, and below 400 m.

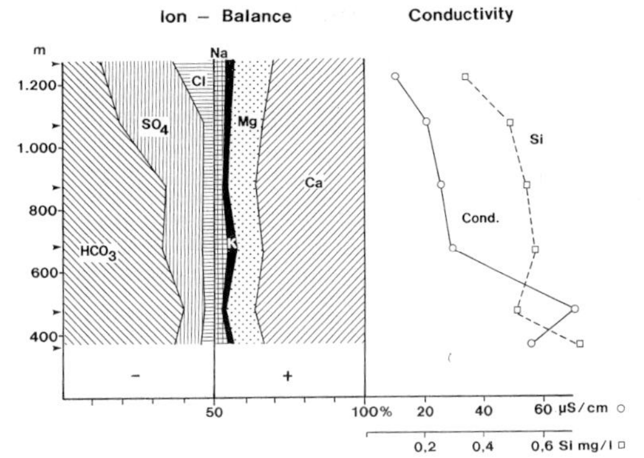

Fig. 3. Average silica content, conductivity, and ionic composition in lakes at different altitude.

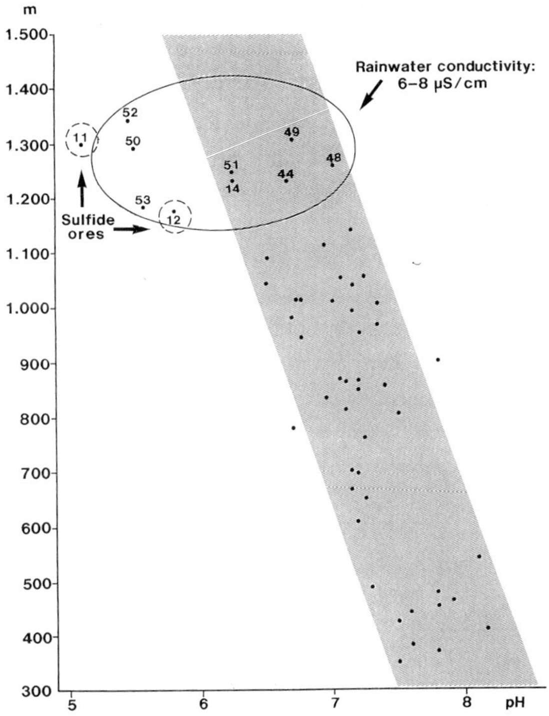

Fig. 4. Lake altitude and pH level.

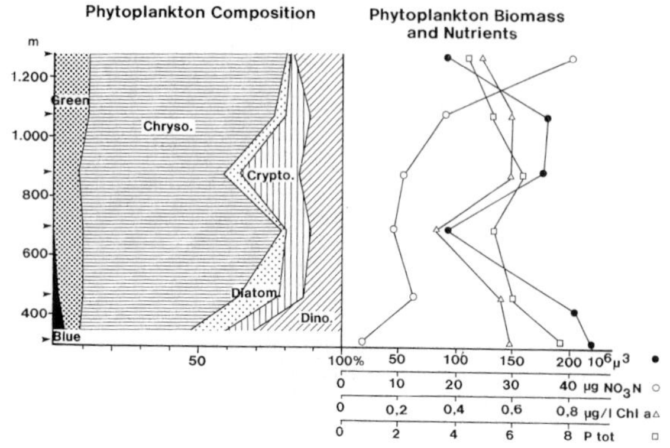

Fig. 5. Average nutrient concentrations (nitrate and total phosphorus), chlorophyll content, algal biomass ($10^6\,\mu^3$ equals roughly 1 mg freshweight), and algal composition in lakes at different altitude (blue = cyanophytes, green = chlorophytes, chryso = chrysomonades, diatom = diatoms, crypto = cryptomonades, dino = dinoflagellates).

nitrogen content. Alpine meadows can contribute organic substances which stimulate productivity even at low phosphorous levels (lakes 32, 33, 38, 55).

Orographic precipitation influences water chemistry

In the Abisko region, main rainfall mainly comes from the north-west (Fig. 1). This can be expected to result in chemical gradients of the lakes' water chemistry. If the lakes are grouped with respect to this main wind direction, the following figures are obtained.

Table 3 shows that orographic gradients exist. These gradients are in accordance with earlier found precipitation chemistry patterns (Nauwerck *et al.*, 1981). There is no doubt that chloride comes with aerosoles from the Norwegian Sea. It could be shown by Nauwerck *et al.* (1981), that even sulphates can come that way and are not necessarily the result of sulphuric acid in the rain. However, poor correlation between chloride and sulphate content in some sulphate enriched lakes rather point to local bedrock as a source of the sulphates than to precipitation.

Table 3. Regional average content of some anions in the lake water.

	meq l^{-1}		
	Na	Cl	SO_4
18 lakes northern part	0.032	0.031	0.114
17 lakes central part	0.015	0.015	0.098
18 lakes southern part	0.011	0.011	0.061

With increasing altitude little change in phytoplankton biomass but change in species numbers and composition

Figure 5 shows that phytoplankton biomass is closely related to chlorophyll *a*, but hardly follows a height gradient. Top lakes certainly show low values, but even lakes in the 600–800 m group have very low values, while values from lakes in the birch region are not much higher than in high mountain lakes of the 1000–1200 m group. Individual differences are large between lakes from all levels. Single very high values like from lake 55, disproportion group averages. The low values in the lakes of the 600–800 m are due to a relatively large number of lakes with high water exchange, which also complicates comparability. Keeping these drawbacks in mind, one may trace a slight biomass decrease with increasing sea level, but it is largely overshadowed by individual differences between single lakes.

More pronounced is the decrease of species numbers with increasing height (Table 2c) and of phytoplankton biomass composition. From nearly 50 species of algae and 9 species of zooplankton in lakes below 400 m they diminish to about 18 species of algae in lakes above 1200 m and about 5–7 species of zooplankton in lakes above 1000 m. Somewhat higher numbers of zooplankton in the top lakes is due to rotifers which may be favoured by the southern exposition – and consequently warmer temperatures – of a majority of those lakes.

As shown by Fig. 5, chrysomonades are the quantitatively most important group throughout the lakes. Their relative importance increases clearly with height. Also green algae show a weak

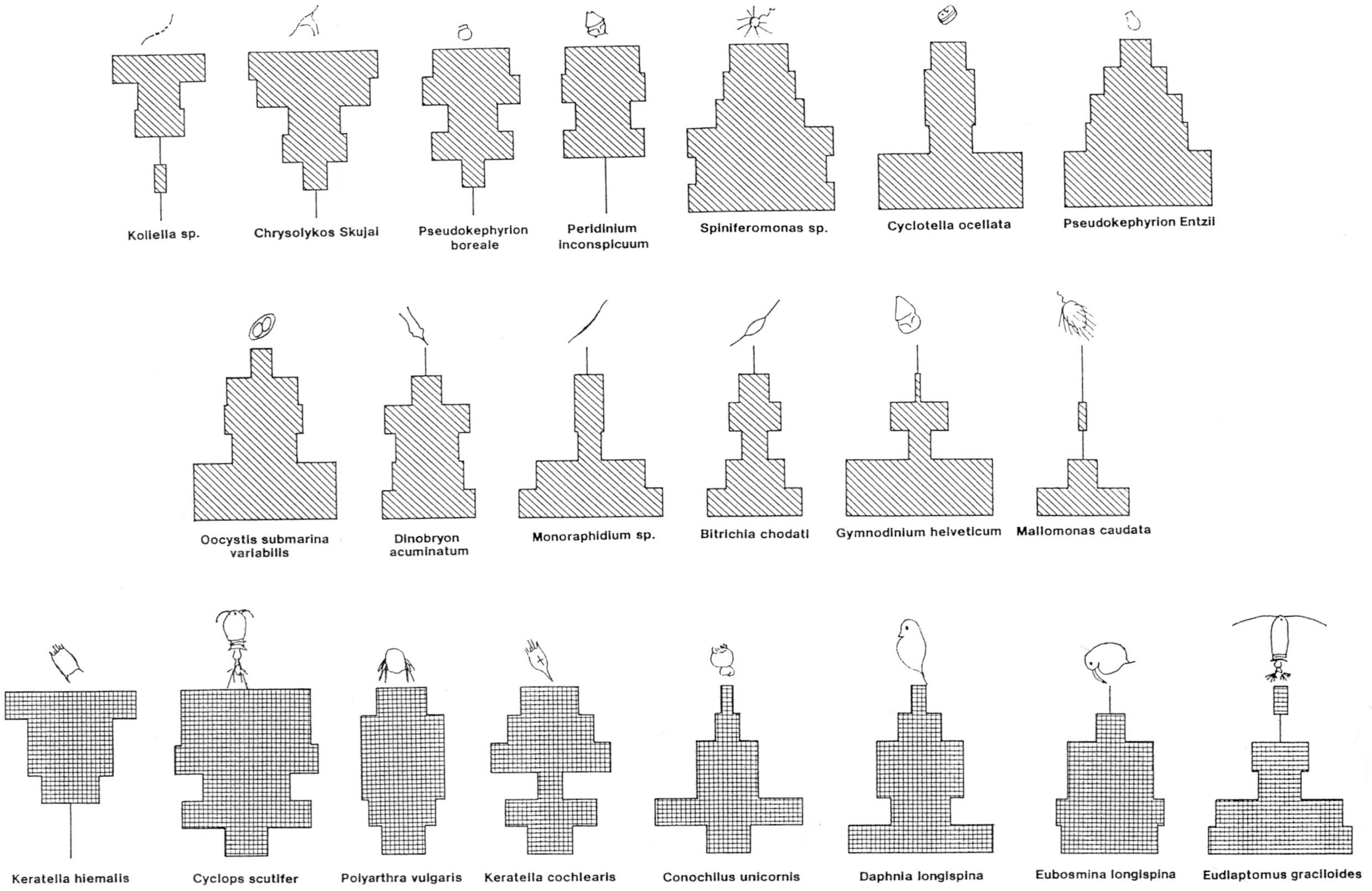

Fig. 6. Frequency of occurrence of some plankton algae and some plankton animals in lakes at different altitude. Grouping of lakes as in Figs 2, 3 and 5.

relative increase and diatoms a weak decrease with heigth. Blue-green algae play a quantitative role only in the lakes of the birch forest region. In these lakes also the diversity of species and the relative contribution of different groups of algae to the total biomass are largest. The relative portion of dinoflagellates remains rather unchanged, but there is a change from mainly large *Gymnodinium* and *Peridinium* species in the lower lakes to small species in the top lakes. This sort of change can be observed even in other groups.

Some plankton organisms show characteristic altitude distribution

In Fig. 6, the relative frequency of some phytoplankton and zooplankton species in different heights is shown. As typical high mountain (or arctic) species among phytoplankton can be noted *Koliella*, *Chrysolykos Skujai* and *Pseudokephyrion boreale*. Even *Peridinium inconspicuum* belongs to the higher lakes. Other species may colonize the highest lakes but are more common in the lower ones, e.g. *Spiniferomonas*, *Oocystis submarina v. variabilis* and others. *Mallomonas caudata* belongs only to the lower lakes. Several other species reach up to the 1000 m mark but are not found in the top lakes. It is remarkable that morphologically similar species in some cases show contrary distributions, like *Koliella*:*Monoraphidium*, and *Pseudokephyrion boreale*:*Pseudokephyrion Entzii*.

Zooplankton distribution is not so clearly stratified as phytoplankton distribution is. Most species are found in at least some of the highest lakes. *Eudiaptomus graciloides* is one species which is most common in the lower lakes, *Keratella hiemalis* is one which is most common in the top lakes. *Cyclops scutifer* and *Keratella hiemalis* appear as the most 'arctic' species. In some of the coldest lakes, the latter is found without posterior spines or with only one. This phenomenon has earlier been described by Nauwerck & Persson (1971). The finding of *Eudiaptomus graciloides* in lake 44 (1228 m) represents a height record of the species. *Daphnia longispina*, in the top lakes often changes into the big *var. rosea* SARS or *var. frigidolimnetica* EKMAN which however, according to Hrbáček (pers. comm.) must be considered as ecotypes of *D. longispina* only.

Table 4. Average limnological properties of 14 Abisko lakes in different summers.

	1968	1981
Conductivity (μS cm^{-1})	32.6	35.8
Water colour (mg Pt l^{-1})	9.5	6.0
Perm. consumption (mg $KMnO_4$ l^{-1})	12.0	4.6
Total phosphorus (μg l^{-1})	6.0	7.7
NO_3-N (μg l^{-1})	8.1	5.9
Secchi depth (m)	9.5	6.0

Interannual meteorological variations superpose possible trends

Table 4 compares average data from 14 lakes of the Abisko region sampled in 1968 by Eriksson & Persson (1971) and by Nauwerck & Ramberg (1979) with averages on the same variables of 1981. The summer of 1968 was exceptionally wet and cool, the summer of 1981 was rather warm and dry. Dilution effects can be traced in 1968, compared to 1981, in conductivity and phosphorous concentrations. Also, in 1968, run-off of humic substances from the surroundings causes

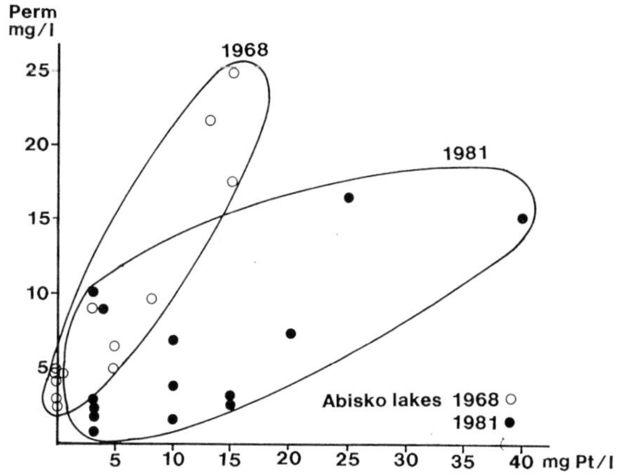

Fig. 7. Comparison between permanganate consumption and water colour in some Abisko lakes 1968 and 1981.

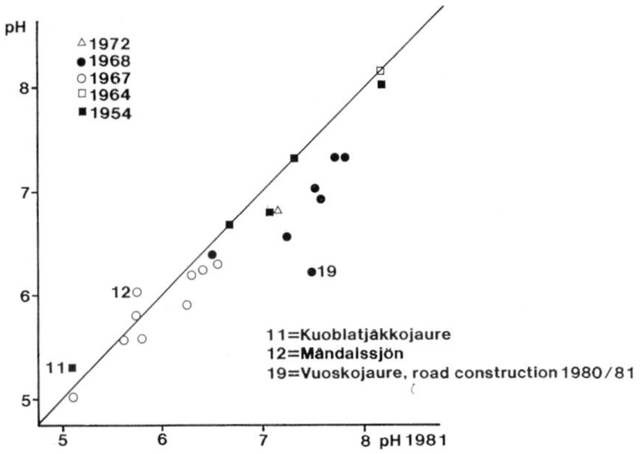

Fig. 8. Comparison between pH values of some Abisko lakes in different years.

higher water colour and, subsequently higher permanganate consumption. Lower productivity results in lower Kjeldahl-nitrogen (a parameter, among other things, for plankton biomass) and, in combination with dilution, increased Secchi depth. Higher productivity in 1981 is also revealed by reduced nitrate values compared to 1968.

Figure 7 shows the relation between water colour and permanganate consumption in the two summers. It can be seen that even the quality of the organic substance which determines permanganate consumption can be different depending on the import of allochthonous material into the lake.

It is important to know that the generally oligotrophic northern lakes usually do not show dramatical variations of chemical properties during the short summer. Therefore, comparisons of single summer data from different years may be allowed and may render at least indicative information. Figure 8 compares pH measurements from different years with the present year 1981. It is obvious that almost every lake in 1981 has higher values than in earlier years. Much higher pH in lake 19 in 1981 certainly was caused by huge amounts of glacial silt which were brought into the water in connection with a road construction. Lower pH, compared to earlier years, in lakes 11 and 12 is to be explained by the withdrawal of permanent snow fields from surrounding rocks and the increasing influence of sulphurous ores.

Conclusions

Northern latitude implies short vegetation period, and, in spite of the midnight sun, low irradiation. With increasing altitude lakes are characterized by

- decrease of temperature, dissolved substances, pH-value, number of plankton species;
- increase of relative share of SO_4, Cl and of chrysomonades; little change of biomass and nutrients;
- special species composition.

Acidification may be indicated by pH values below 5.5 in some lakes. Local particularities and short term meteorological effects involve a wide range of variation and offers natural explanations. There is no measurable proof for acidification.

References

Ahl, T., 1972. Hydrochemical analyses. A catalogue of procedures used within the IHD in the Nordic countries. Nordic IHD Report No. 3, 69 pp.

Almer, B., W. Dickson, C. Ekström, E. Hörnström & U. Miller, 1974. Effects of acidification on Swedish lakes. Ambio 3: 330–336.

Anderson, R. S., 1971. Crustacean plankton of 146 alpine and subalpine lakes in western Canada. J. Fish. Res. Bd Can. 28: 311–321.

Caplanqc, J., 1972. Phytoplancton et productivité primaire de quelques lacs d'altitude dans les Pyrenées. Ann. Limnol. 8: 231–321.

Ekman, S., 1904. Die Phyllopoden, Cladoceren und freilebenden Copepoden der nordschwedischen Hochgebirge. Zool. Jahrbücher, 21, Abt, f. System.

Ekman, S., 1957. Die Gewässer des Abisko-Gebietes und ihre Bedingungen. Kungl. Sv. Vetenskapsakad. Handl., ser. IV, 2:1, 3:3, 4:1, 4:5, 5:4. Stockholm.

Ekman, S., G. Lohammar, W. Rodhe och H. Skuja, 1950. Undersökningar av sjöar i Torne Lappmark med särskilt hänsyn till deras plankton och vattenkemi. Statens Naturvetenskapliga Forskningsrådets Årsbok 1948–1949. Stockholm.

Eriksson, G. & G. Persson, 1971. Limnologiska studier i högfjällsvatten i Latnjajaureområdet 1967. Scr. Limnol. Uppsal., Collect. 7A, Ser. 278.

Holmgren, S., 1968. Phytoplankton production in a lake north of the arctic circle. Fil.lic.-thesis, Limnol. inst. Uppsala, 39 pp.

Honsig-Ehrenburg, W. & R. Psenner, 1986. Zur Frage der

Versauerung von Hochgebirgsseen in Kärnten. Carinthia II, 176/96: 443–461.

Jansson, M., 1972–1978. Hydrologi och vattenkemi i: Kuokkelprojektet, Experiment med sjögödslingar i Kuokkelområdet. Årsrapport I–VI. Stencil, Limnol. inst. Uppsala.

Larson, G. G., 1973. A limnological study of a high mountain lake in Mount Rainier National Park, Washington State, USA. Arch. Hydrobiol. 72: 10–48.

Mosello, R., 1986. Effects of acid depositions on subalpine and alpine lakes in NW Italy. Mem. Ist. ital. Idrobiol. 74: 117–146.

Nauwerck, A., 1966. Beobachtungen über das Phytoplankton klarer Hochgebirgsseen. Schweiz. Z. Hydrol. 28: 4–28.

Nauwerck, A., 1968. Das Phytoplankton des Latnjajaure 1954–1965. Schweiz. Z. Hydrol. 30: 188–216.

Nauwerck, A.,1978. Bosmina obtusirostris SARS im Latnjajaure. Arch. Hydrobiol. 82: 387–418.

Nauwerck, A., 1980. Die pelagische Primärproduktion im Latnjajaure, Schwedisch Lappland. Arch. Hydrobiol/ Suppl. 57: 291–323.

Nauwerck, A., 1981. Surhetsförhållandena i ytvatten i Norrbottens län. Länsstyrelsens i Norbbottens läns rapportserie 1981: 7, 18 pp.

Nauwerck, A., 1983. Vattenkemi och plankton i sjöar i Abiskoområdet augusti 1981. Länsstyrelsens i Norrbottens läns rapportserie 1983: 14, 26 pp.

Nauwerck, A. & G. Persson, 1971. Ekmanjaure – en återfödd sjö. Fauna och flora 66: 130–140.

Nauwerck, A. & L. Ramberg, 1979. En regional studie över fytoplankton och vattenkemi i Abiskoområdet. Länsstyrelsen i Norrbottens läns Rapportserie 1979: 11, 23 pp.

Nauwerck, A., L. Lindquist & H. Groth, 1981. En nederbördskemisk snöinventering i Norrbottens län mars 1980. (Snow precipitation chemistry in the county of Norrbotten, March 1980). Länsstyrelsens i Norrbottens läns rapportserie 1981: 11, 33 pp.

Ohle, W., 1940. Chemische Gewässererkundung in Schwedisch-Lappland. Arch. f. Hydrobiol. 36: 337–358.

Pejler, B., 1957. Taxonomical and ecological studies on plankton rotatoria from northern Swedish Lapland. Kungl.Sv. Vetenskapsakad. Handl., ser. IV, 2:1: 1–68.

Psenner, R., U. Nickus & F. Zapf, 1988. Versauerung von Hochgebirgsseen in kristallinen Einzugsgebieten Tirols und Kärntens. Zustand, Ursachen, Auswirkungen, Entwicklung. Projektrapport des BMLF, Wien, 335 pp.

Raddum, G., A. Hobaek, E. Lomsland & T. Johnson, 1980. Phytoplankton and zooplankton in acidified lakes in southern Norway. In Proc. Int. Conf. Ecol. Impact of Acidification. SNFS projekt, Oslo-Aas, Norway: 332–333.

Rodhe, W., 1963. Livet i högfjällssjöarna. Natur i Lappland. K. Curry-Lindahl ed., Bokförlaget Svensk Natur, Uppsala.

Rodhe, W., J. E. Hobbie & R. T. Wright, 1966. Phototrophy and heterotrophy in high mountain lakes. Verh. int. Ver. Limnol. 16: 302–313.

Skuja, H., 1964. Grundzüge der Algenflora und Algenvegetation der Fjeldgegenden um Abisko in Schwedisch-Lappland. Nova Acta Reg. Soc. Sci. Upsal., ser. IV, 18:3, 465 pp.

Hydrobiologia **274**: 101–114, 1994.
J. Fott (ed.), Limnology of Mountain Lakes.

Cyclops scutifer SARS in Lake Latnjajaure, Swedish Lapland

Arnold Nauwerck
Limnological Institute, Austrian Academy of Sciences, A-5310 Mondsee, Austria

Key words: Cyclops scutifer, Lapland, distribution patterns, population dynamics

Abstract

In Latnjajaure, an ultra-oligotrophic mountain lake in Swedish Lapland, *Cyclops scutifer* is always present in considerable numbers. During a developing time of 3 years from egg to adult, losses remain low. High survival rates may be explained by lack of predators, even if there is selfcontrol by cannibalism. Temperature, not food supply, appears to be the most important growth steering factor. During more than 9 months of winter, the population is accumulated in water layers below 15 m depth. After the ice-break (July), the adults display diurnal vertical migration, and the distribution pattern in the lake appears extremely patchy. Food search and mating advantages are possible reasons for this behaviour.

Introduction

Lake Latnjajaure in Swedish Lapland is an ultra-oligotrophic mountain lake. The lake has no fish. A broad study of the lake ecosystem was started in 1964 (Nauwerck, 1967) and was completed at the end of the sixties. Because of outward circumstances, the long-term objective of the project, to study the changes of the system after the introduction of fish, was so far not realized. In the meantime, fish-free lakes have become rare in the Scandinavian mountains. Therefore it seems more sensitive to put the lake under protection in its original state than to go on with the project.

Lake Latnjajaure is situated in the western mountains of the Abisko area in northern Swedish Lapland. Its height at 987 m above sea level corresponds phytogeographically and climatically to roughly 3000 m in the Alps. Its surface covers 0.73 km^2 which makes about one tenth of the total drainage area. The theoretical time for the water exchange of the lake is 4 years. The lake is extremely clear (maximal Secchi depth 35 m) and a moss carpet of *Marsupella aquatica* (SCHRADER) SCHIFFNER stretches down to almost 40 m depth. Ice break occurs usually in July, and a new ice cover is formed in October. Low temperature, short vegetation period and very low nutrient content (average total phosphorus content $<4\ \mu g\ l^{-1}$) result in extremely low primary production (annual mean 2.7 g C m^{-2}). Nevertheless, a considerable standing stock of zooplankton is permanently present in the lake.

There is a considerable amount of information on Lake Latnjajaure. Early data stem from Ekman (1904, 1957), Pejler (1957), Rodhe (1962, 1963), Lohammar (1963), Nauwerck (1966), Rodhe *et al.* (1967). Results of the Latnjajaure project have been published by Taube & Nauwerck (1967), Nauwerck (1968a, 1978, 1980, 1981), Bodin & Nauwerck (1968), Wåhlin (1970), Hellström & Nauwerck (1971), Nauwerck & Persson (1971), Schönborn (1973, 1975). Further, there are technical reports by Palm & Törmä

(1965), Bodin (1966), Taube (1966), Lithner (1968), Nauwerck (1968b) & Eriksson & Persson (1971).

In Lake Latnjajaure *Cyclops scutifer* SARS is the most important zooplankton species beside *Eubosmina longispina* (LEYDIG) (= *Bosmina obtusirostris* SARS). *Cyclops scutifer* is also the most common zooplankton species in the mountain lakes of the Abisko region (Ekman, 1904; Nauwerck, 1983), and, together with *Keratella hiemalis* CARLIN, it colonizes even the uppermost high alpine lakes of the area. In many lakes where it lives, *Cyclops scutifer* faces very low temperatures and very low food supply. Prolongated development time is one possibility to elude the shortcomings of such an environment, which obviously is not chosen by the animal because of preference but rather because of lack of competition (Taube & Nauwerck, 1967). McLaren (1961) has reported a biennial population of *Cyclops scutifer* in Lake Hazen, Ellismere Island. Even in Norway extraordinary prolongation of the life cycle of the species has been stated (Elgmork, 1965, 1981; Elgmork & Eie, 1989). Coexisting populations of different morphs of the species which have been observed in some lakes of the Scandian mountain chain (Lindström, 1958; Axelson, 1961) may be explained as a mixture of one year and two year cycles in one and the same lake.

Along with other properties of Lake Latnjajaure, its zooplankton has been studied quantitatively from 1964 to 1969. Results concerning *Eubosmina* have been published (Nauwerck, 1978) together with a few quantitative notes on *Cyclops*.

Other members of the zooplankton of the lake are *Keratella hiemalis* CARLIN, *Kellicottia longispina* (KELLICOTT) and *Polyarthra vulgaris* CARLIN. These rotifers are relatively abundant. *Filinia terminalis* PLATE is found occasionally. *Megacyclops gigas* (CLAUS), *Canthocamptus arcticus* LILLJEBORG und *Eurycercus lamellatus v. frigida* EKMAN live mainly close to the bottom. *Eudiaptomus graciloides*, *Daphnia longispina var. rosea* SARS (= *var. frigidolimnetica* EKMAN) and *Chydorus sphaericus* (O. F. MÜLLER) were reported, among others, by Ekman (1904), *Keratella cochlearis* (GOSSE) by Pejler (1957). They appeared up to the beginning of the sixties but were not found any more during the period of the Latnjajaure project. A series of warm years seems to be necessary for them to spread in the lake.

Material and methods

At the time this project was run, field work had to be carried out under primitive conditions and with simple equipment. Though it soon became evident that sampling volumes of less than 10 liters of water were not sufficient, from a statistical point of view, for quantitative zooplankton work in the lake, there was no chance to do much better. Sampler size and sampling time were limiting factors. During the first years samples of 9 liters of water (5 Ruttner bottles of 1.8 liters) were taken from different depths at one standard point. The water then was filtered through 100 μm gauze, and the remaining animals were washed down into a bottle and fixed by formaline for later counting under an inverted microscope. In later years sampling volumes were increased to 20 liters and even 50 liters with the help of a sunken battery driven pump. Comparisons showed good correspondence with bottle samples of increasing size.

Nevertheless, the quantitative samples fulfilled neither the requirements for calculations of population dynamics nor of production. The real obstacle is the heterogeneity of plankton distribution. For one, the horizontal distribution turned out to be extremely patchy (cf. Figs 2 and 3). For the other, an increase of total numbers of animals can be observed at times for other reasons than reproduction (see below). Moreover, diurnal vertical migrations could result in enormous accumulations of *Cyclops*, particularly males, at the water surface. Thus, on one occasion (3rd of August 1967), over a depth of 20 m, *ca* 200 Cyclops males and some females were found per liter in the very surface water, but just a few individuals per liter in only 10 cm below.

Population dynamic calculations therefore had to be based on net samples, which were taken 'semi-quantitatively', which means a vertical stretch of 30 m at a standard point was passed 6 times with a 100 μm plankton net. Equivalent subsamples were taken from these samples and all *Cyclops* individuals, including nauplii 300–1000 animals, were measured and were classed according to developing stages. This allows a better analysis of the population dynamics than the quantitative counts do and also allows relative comparisons of total amounts.

Results

Temporal distribution

With respect to the above mentioned problems, quantitative data from single years do not reflect real interannual differences. Therefore, only the average annual distribution pattern is demonstrated (Fig. 1). It is a mean of the data of 6 years, altogether 49 sampling occasions, samples taken from 3–5 depths.

The average number of individuals fluctuates between about 7 l^{-1} during the season of ice cover and darkness, and about 15 l^{-1} in July. The increase, still under ice, at the beginning of the light season, can only be explained by remigration of the animals from refuges close to the bottom into the open water. Nauplii count for more than 50% of the population most of the year. Their portion minimizes in August to about 25% of the population. Adults are found practically only during the short summer season from July until the beginning of October and make up little more than 15% of the population, at the most 2.5 ind l^{-1} in July.

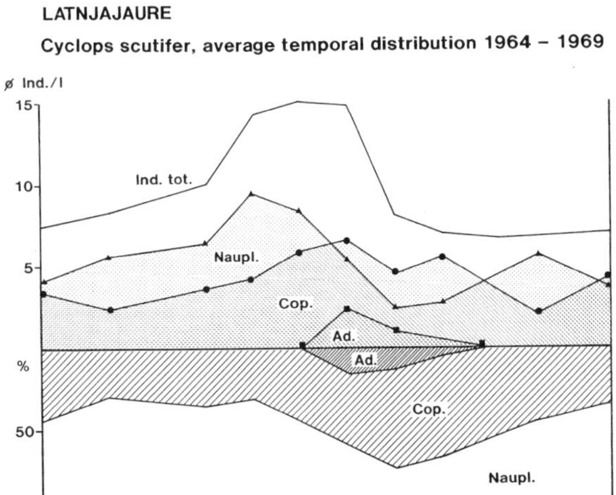

Fig. 1. Latnjajaure. Average temporal distribution and age composition of Cyclops scutifer, 1964–1969.

Horizontal distribution

During the summer of 1968, hatching traps for chironomids were set out at 78 places all over the lake and were checked at *ca* 10 day intervals. In these plastic funnel traps, placed at 2–3 m depth, also the vertically migrating *Eubosmina* and *Cyclops* were caught. In neither case is there any relation between the number of caught animals and actual water depth. It is not possible to translate the catches into absolute numbers per water volume, but they can be used as a measure of the evenness or unevenness of horizontal distribution of the animals.

In contrast to *Eubosmina*, which showed large scale agglomerations of surprising temporal stability in certain parts of the lake (Nauwerck, 1978), *Cyclops* copepodides display a more patchy picture which changes strongly from time to time (Fig. 2). The two examples from August and September show density differences within a short distance like 1:1000, and differences of similar magnitude at one and the same spot between two sampling occasions.

Horizontal and vertical distribution were checked along with this trapping on a number of profiles from close to the shore to the deep parts of the lake (Fig. 3). While a remarkable unevenness can be stated in August, with respect to horizontal as well as to vertical distribution, vertical distribution pictures are more even in September. One is under the impression that the animals now withdraw to deeper water layers.

Vertical distribution

Normally, sampling was done at daytime. Figure 4 shows the average depth distribution of *Cy-*

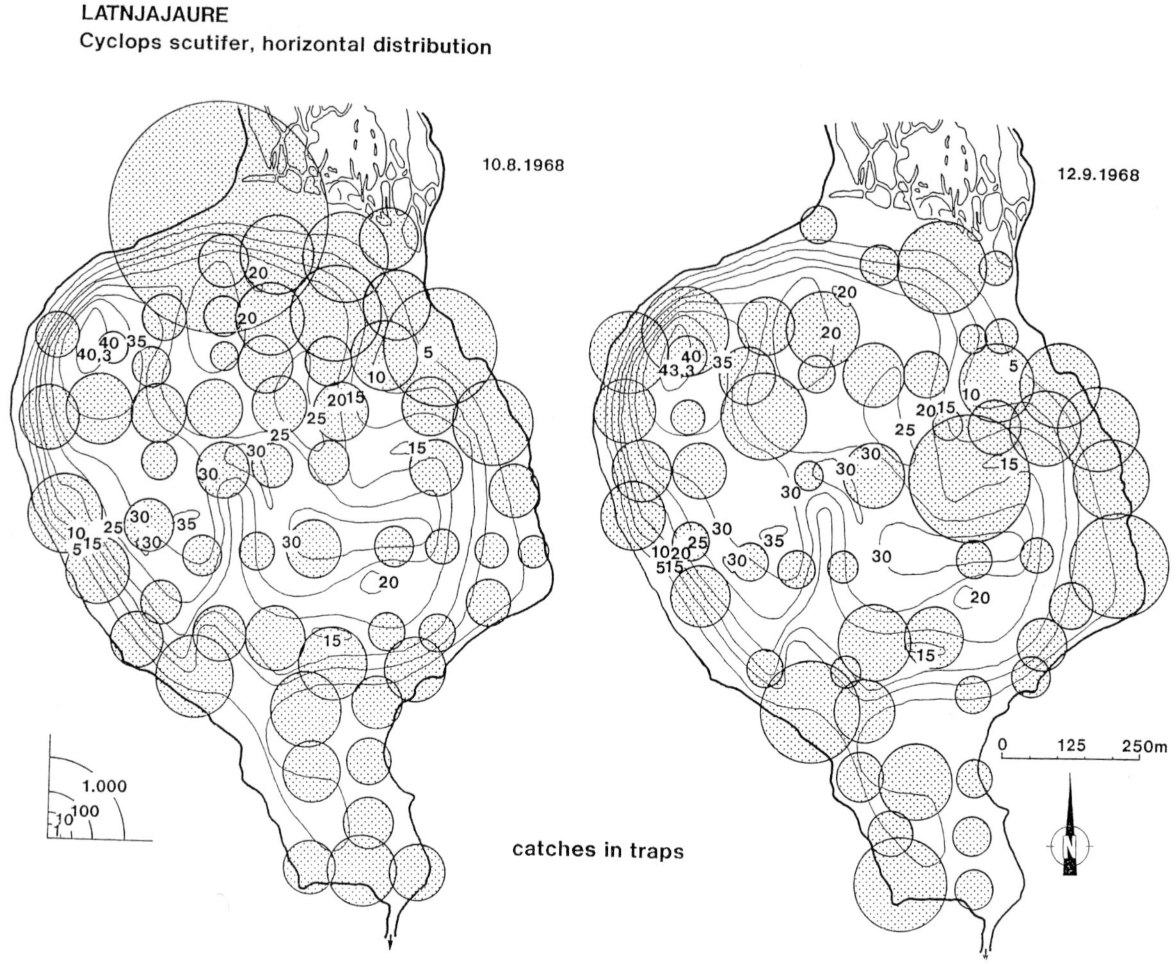

Fig. 2. Latnjajaure. Horizontal distribution of Cyclops scutifer. Trap catches, 1 August–10 August and 3 September–12 September 1968.

clops throughout the year. The thick line in Fig. 4 indicates the population balance point of copepodides and adults. Half the population is found above and half the population below this depth. The stippled line gives the same for the nauplii. The hatched area includes 50% of the population.

As can be seen, the population balance point is below 20 m depth for nauplii all the time, while the more advanced stages are found in higher water strata during spring when light increases under the ice cover.

Figure 5 shows diurnal distribution patterns at different times of the year. Adults show strong diurnal migration in summer, males' migration amplitudes seem to be larger than that of females. Some upward migration during night and sinking or downward migration during day can be noted in the copepodides (hatched area). It seems to be more pronounced in fall than in summer. Nauplii do practically not migrate.

Population dynamics

The composition of different stages of *Cyclops* in the net samples is shown in Fig. 6 for the last three years of the study. At a first glance one can see that nauplii and the copepodide stages III and V are predominant during the long winter season

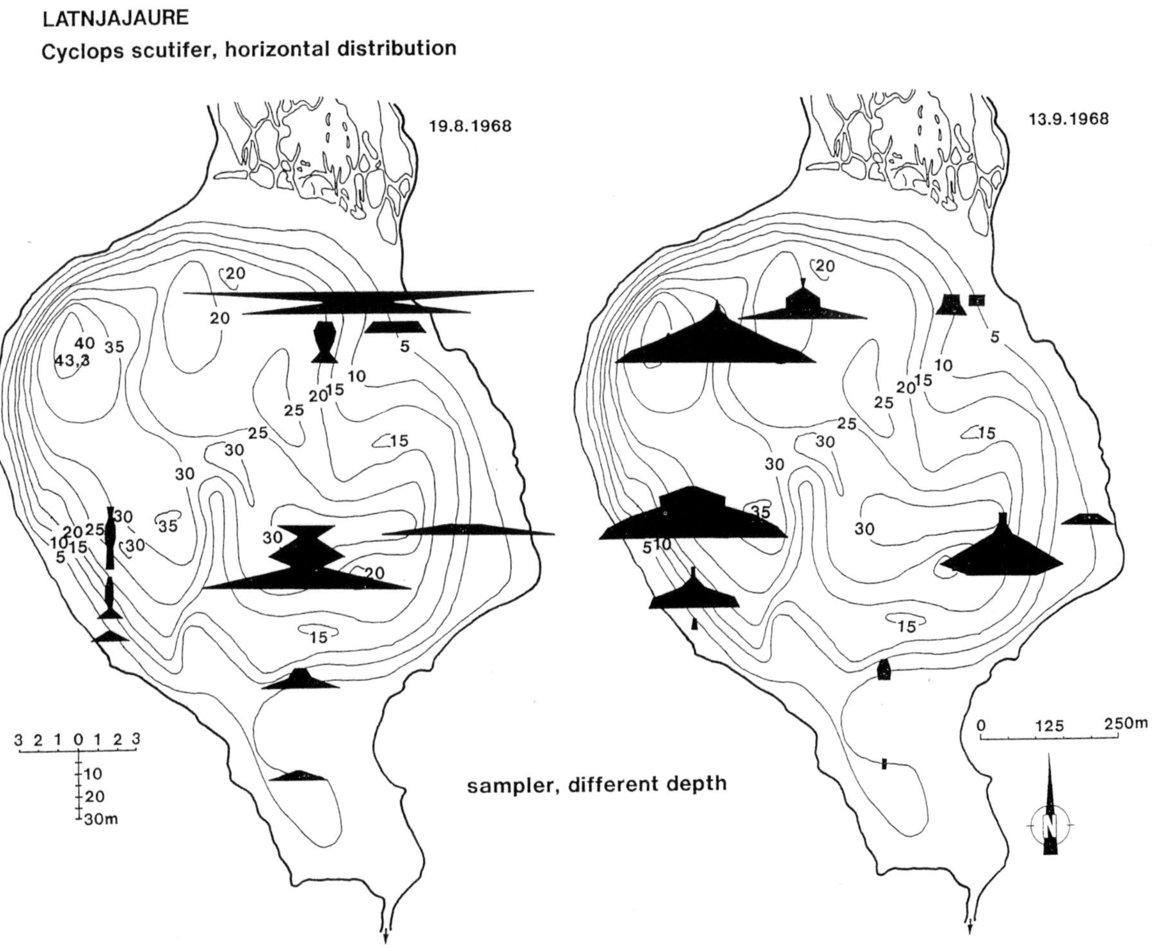

Fig. 3. Latnjajaure. Horizontal and vertical distribution of Cyclops scutifer along some depth gradients 19 August and 13 September 1968.

when otherwise little happens. Almost all changes take place during the short summer. Nauplii change quickly into copepodides I and II, and for a short time there are almost no nauplii left. The copepodides also continue quickly to copepodide stage III.

At the same time, present copepodides IV switch into copepodides V (or into males) and earlier copepodides V switch into females. The adults disappear in September–October. Obviously the switch from copepodide I to copepodide II takes place only during summer, as well as the switch from late copepodides to adults, but the switch from copepodide II to copepodide III and from copepodide IV to copepodide V may drag on for a longer period of time.

As shown by Fig. 6, males always hatch earlier than females and die out, more or less, at the end

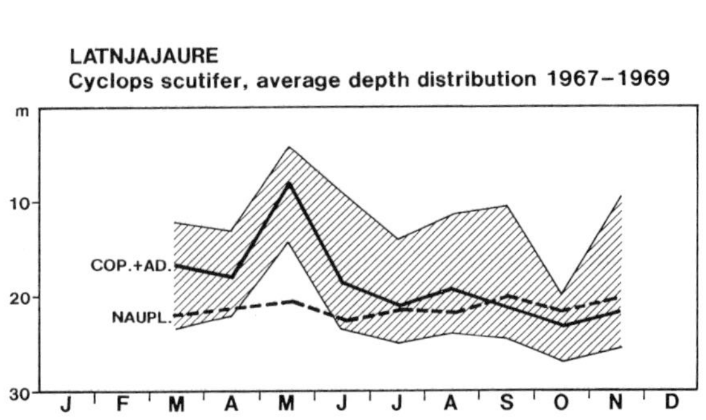

Fig. 4. Latnjajaure. Average depth distribution 1967–1969. Shadowed area is 50% of total population. Fat line and stippled lines show 50% percentiles.

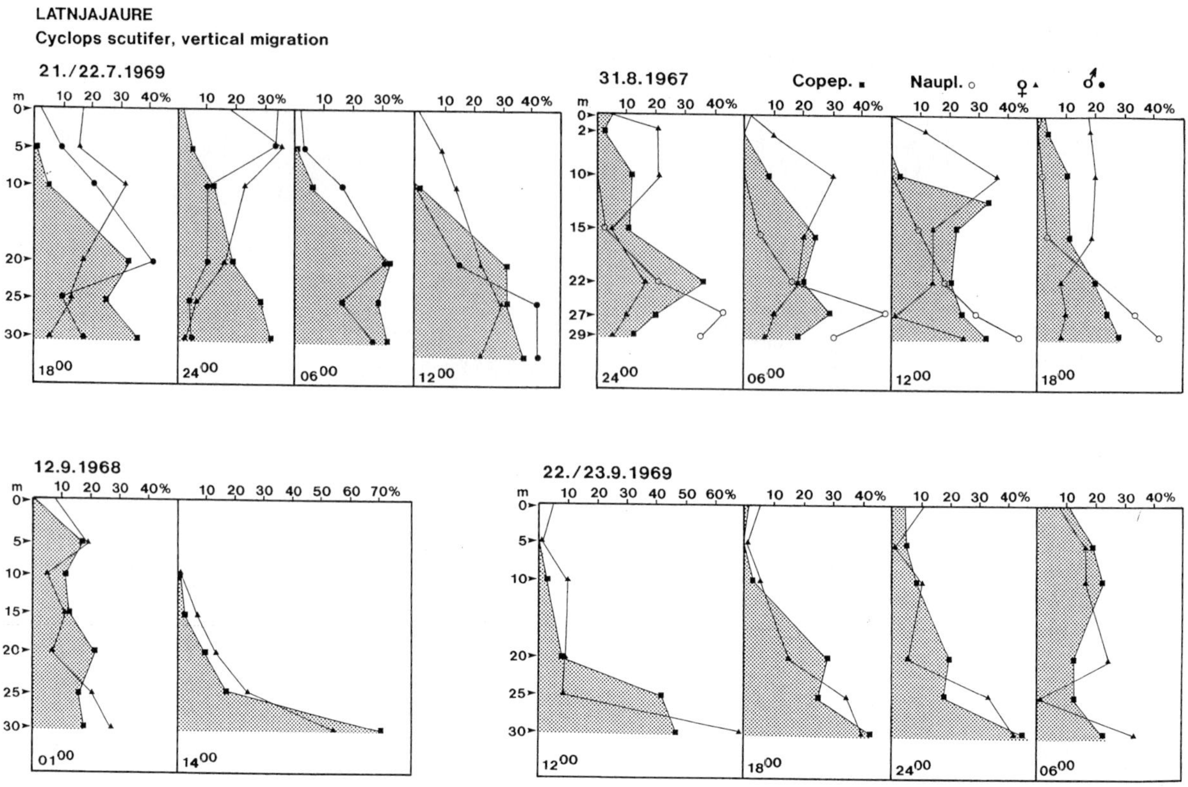

Fig. 5. Latnjajaure. Vertical migration of Cyclops scutifer at different occasions. Shadowed area shows copepodides.

of August. As to be seen in Fig. 7, a 50:50 level of the sexes is established towards the end of July in 1964, 1965, 1967 and 1968. In 1966 this point seems to be delayed and in 1968 is appears earlier. Proceding cold or warm years respectively (see Fig. 9) may be taken into consideration to explain the deviations.

Production and losses

The period of egg production (Fig. 8) reaches from the middle of July until the end of September. Clutch sizes are largest at the beginning and successively diminish towards the end of the season. Highest average numbers are around 20. In the cold summer of 1965, clutch sizes are distinctly smaller than in other years. There is also a delay of almost one month compared to other years. In the warm summer of 1967, a second peak of egg-laying females may indicate that hibernating copepodide IV could pass through copepode V and reach adult stage the very same summer. Lack of a corresponding increase of males, however, contradicts such a probability. The favorable conditions of that summer may have allowed more females than usual to produce a second, minor clutch of eggs.

The nauplii hatch from August until October (Fig. 6). A decrease of numbers which is not compensated by a corresponding increase in the next stage can be observed in the nauplii, particularly in the summer of 1968, and in the copepodides II and V during the winter of 1967/68. This decrease may indicate losses as well as a move towards the sediments. An increase of the population because of remigration from the bottom is marked in the nauplii and in copepodide III in June–July.

Losses from nauplii up to adults is in an order of magnitude of 90% (*ca* 50 survivors out of 500 nauplii). The losses between nauplii and copepodide I can be estimated to 20–50%, the losses

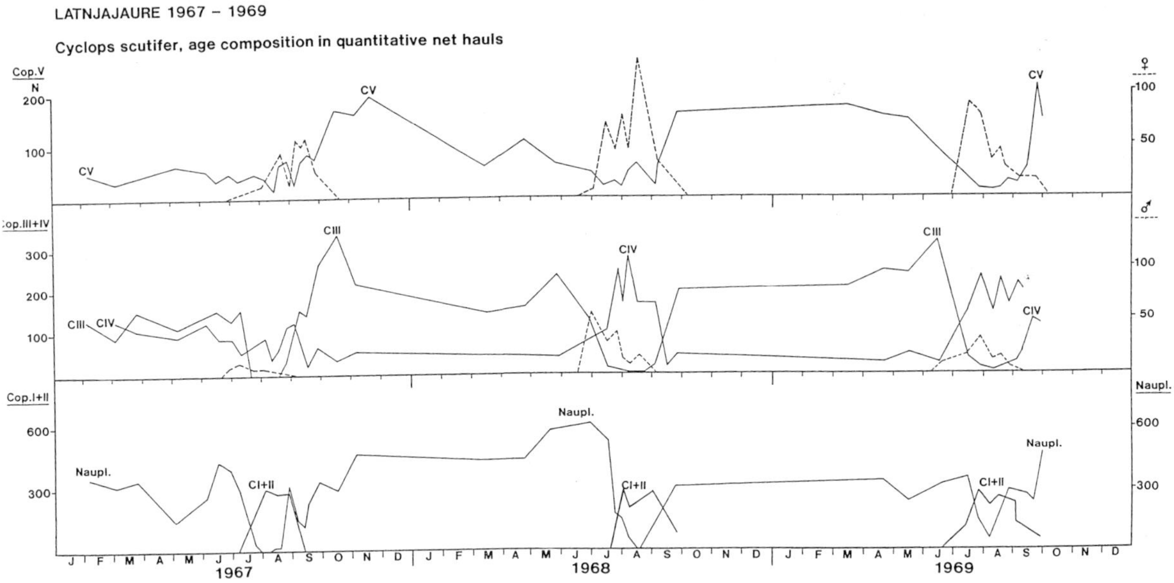

Fig. 6. Latnjajaure. Age composition of Cyclops scutifer in semi-quantitative net in semi-quantitative net hauls.

between copepodide III and IV maybe 30% and the losses between copepodide V and adult another 50% of the numbers of the preceding stage. Only minor losses occur between copepodide stages I and II and between stages IV and V.

Temperature as a steering factor for population dynamics and growth

Figure 9 combines age composition of the *Cyclops scutifer* population at different times with average lake temperature and phytoplankton biomass from 1965 to 1969. The time scale for the summer months is stretched. The drain from nauplii to adult is marked as shadowed areas.

A triennal development can be distinguished. Young *Cyclops* are born in August and September. They move up to higher nauplius stages relatively quickly through September and only very slowly during the following 10 months of winter. In July of the following year, they pass quickly through copepode I and II stages and stay as copepode III during next winter. The second summer they go on to copepode IV and V, and again more or less stop to develop during the third winter. It seems that the overwintering copepodide IV is the one to switch into males at the beginning of the third summer while the copepod V turns into females.

During the years of this study, Lake Latnjajaure went through a period of increasing temperatures. However, the increase of summer temperatures, whether it was fast as in 1966 or slow as in 1968, influences the development of the *Cyclops* population less than does an increased winter temperature. It makes a considerable difference if the living temperature of the animals for 10 months is 1 °C or if it is 2 °C. Cold winters 1964–1965 and 1965–1966 thus are the reason for the delay in the development more than actual summer temperatures. The relatively warm summer of 1966 cannot compensate for the winter retardation, and the relatively cold summer of 1968 cannot essentially brake the development of a preceding warm winter.

Figure 10 shows that the average size of the nauplii – in this case also a measure for stage composition – is clearly linked to lake temperature. The warmest preceding winter of 1967 results in the largest/most developed nauplii, the warmest summer of 1969 results in fastest growth, the coldest summer of 1968 in slowest growth. Also the size of single copepodide stages is linked

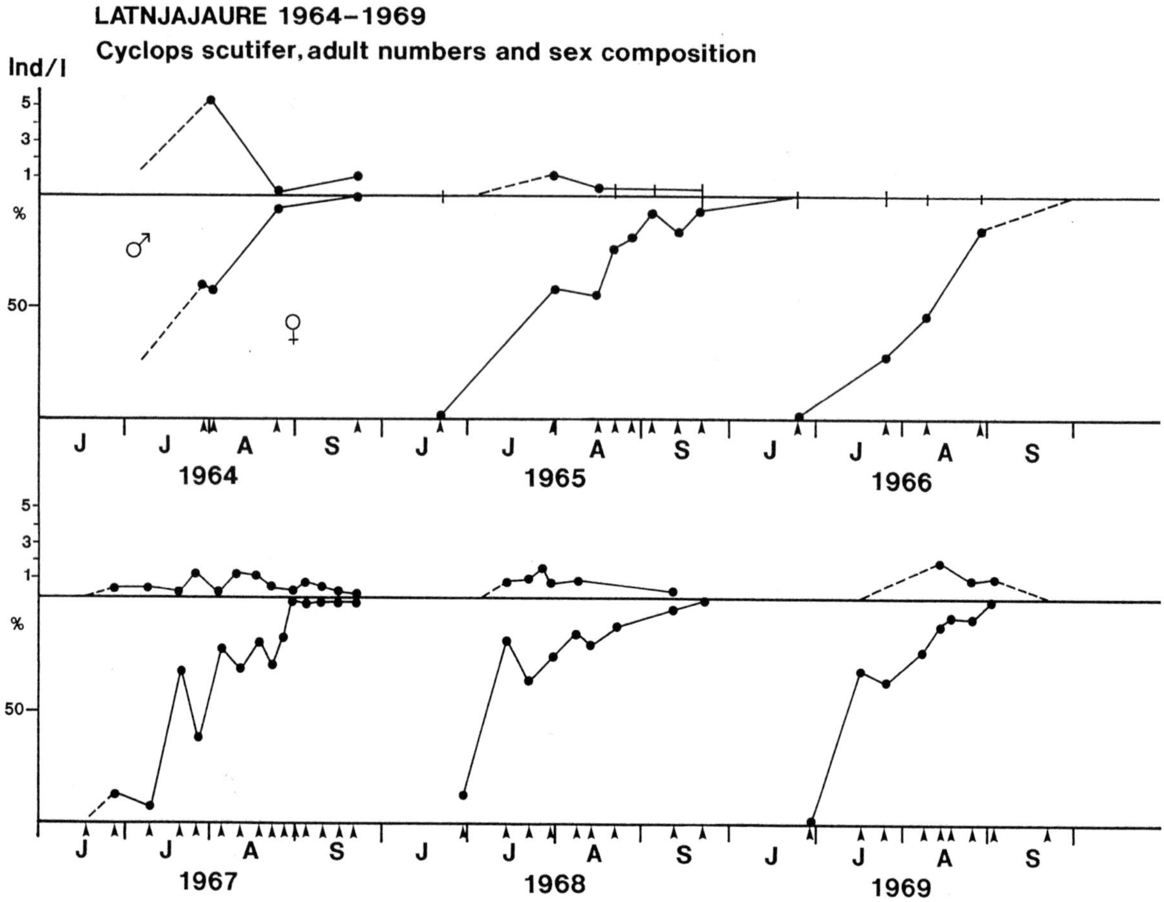

Fig. 7. Latnjajaure. Adult numbers and sex composition 1964–1969.

to temperature. Particulary remarkable is the difference between the size of young copepodide stages of the same age from different summers.

Food as a steering factor of population dynamics and growth

While temperature certainly is the main driving force for physiological processes, and thus for the developing speed of our *Cyclops*, food supply must be taken into account as a modifying factor, whose influence is greater, the longer the animal remains at a certain developing stage and the older it grows.

The nauplii are known to live on yolk reserves for a couple of stages and start feeding on small particles as metanauplii. Suitable particles are small algae and bacteria. According to our observations, young copepodides too feed on small particles. With increasing age the animals become omnivorous and also eat their own offspring. In gut content and fecal pellets, different algae like diatoms, small gymnodineans, green algae, *Dinobryon* loricas as well as bottom living algae were found along with remains of crustaceans. A comparison of *Cyclops*' development with the supply of phytoplankton thus may add to the understanding of interannual differences in growth and productivity of the animals.

Figure 9 gives the phytoplankton development in different years (see also Nauwerck, 1968a, 1980). The year 1965 was extremely unproductive. The phytoplankton biomass was domi-

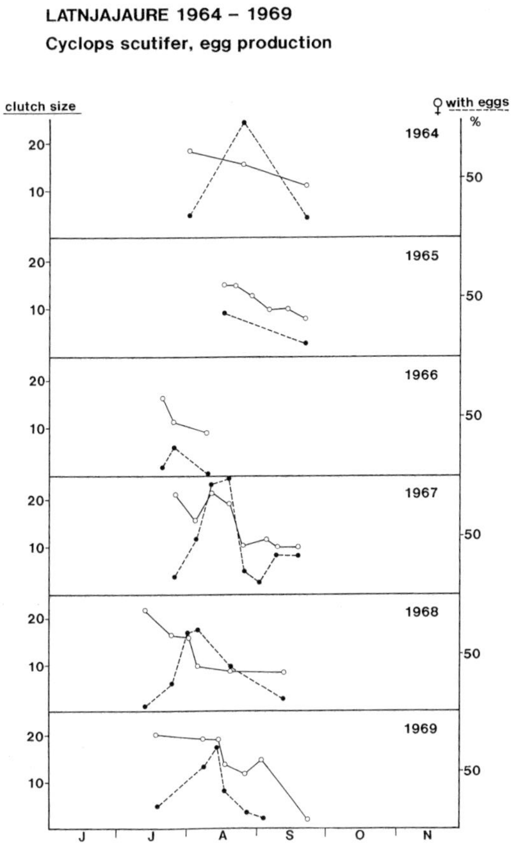

Fig. 8. Latnjajaure. Cyclops scutifer, egg numbers 1964–1969.

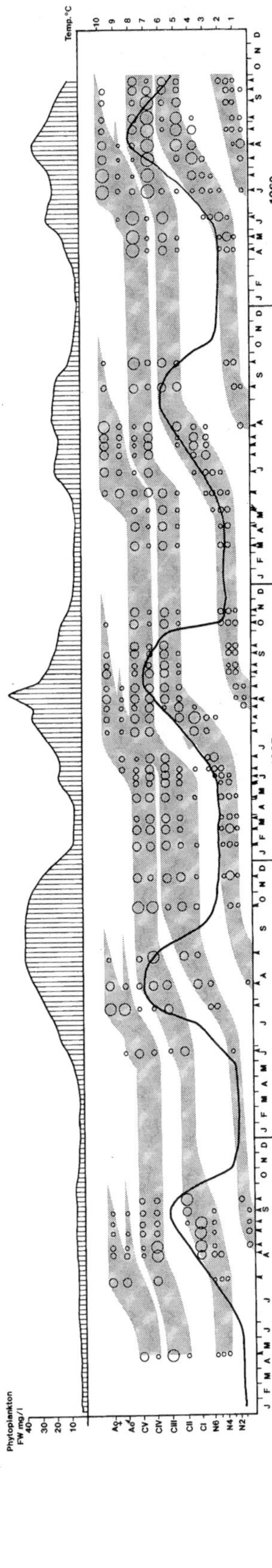

Fig. 9. Latnjajaure 1965–1969. Cyclops scutifer population dynamics compared to phytoplankton biomass and temperature.

nated by small *Peridinium inconspicuum*. This species can be grasped by late copepodites and adults of *Cyclops* but not by younger stages. In 1966, phytoplankton biomass development was high, but the dominating species was *Oocystis submarina v. variabilis*, a green alga with relatively thick cell walls. It could be found that it usually passed the gut of the animals undigested and obviously had little food value. In 1967 there was a richness of small chrysomonades, a very suitable food source for phytophagous zooplankton. Also in 1968 the supply of chrysomonades was good, but the phytoplankton biomass included a larger portion of dinoflagellates. In 1969 small diatoms like *Cyclotella glomerata* together with chrysomonades, again offered favorable food conditions.

Even if not strongly supported by the data, there is a direct positive correlation between food

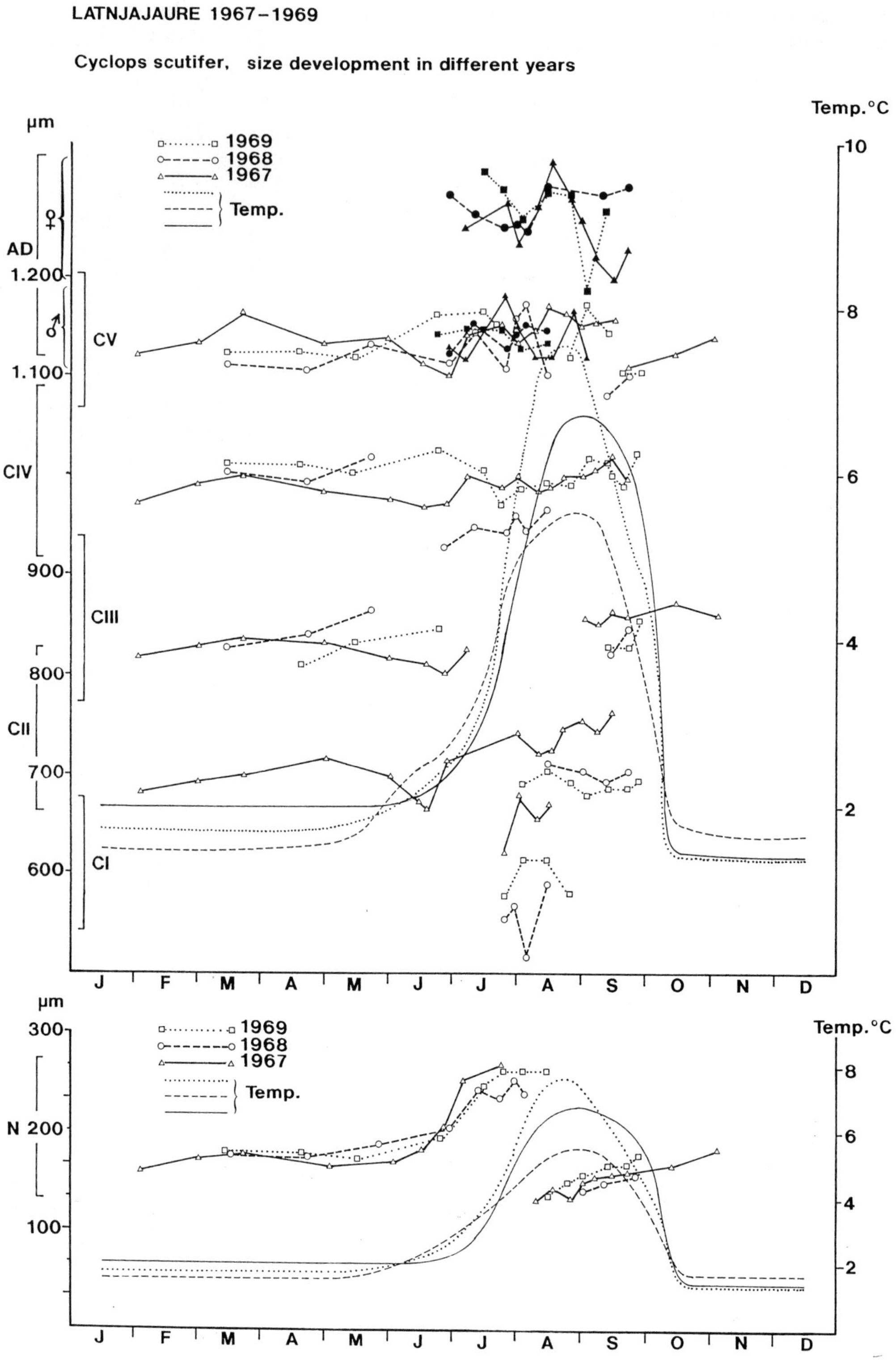

Fig. 10. Latnjajaure. Cyclops scutifer nauplii, copepodide and adult size development (body length) and temperature 1967–1969.

supply and average egg numbers during summer. Highest average egg numbers can be estimated in 1967 and 1969 (15.0), lower numbers are found in 1965 and 1968 (12,5) and the lowest numbers in 1966 (10,0), also 1964 shows high numbers. Such a direct correlation with actual food supply is a bit surprising. Accumulated nutrient reserves from a long copepodide existence are more likely to grant steady reproduction under generally unfavorable conditions than does an immediate response on occasionally improved food conditions. So far as total egg production can be estimated, on the other hand, 1966 is a good year and both 1967 and 1968 rather show medium values. However, a previous life story of three winters and two summers may have influenced growth, losses and condition of the animals in several ways.

Figure 11 shows, schematically, the changing conditions different generations of *Cyclops* had to face during their life. (Information from before 1964 from Rodhe, 1962, 1963; Rodhe *et al.*, 1966, also unpublished data.) The generation which was born in 1961 and reached adulthood in 1964, experienced favorable conditions with respect to both food and temperature through its whole life. The animals born in 1962 and adult in 1965 experienced low winter temperatures twice, as copepodide III and V, ended up as adults in a very poor year and obviously also suffered high losses. Animals born in 1963, adult in 1966, also had a good start into life but had to endure two cold winters and one very unfavorable summer. As the age-class was numerous, many of them reach adulthood anyhow, and total egg production was high, but individual productivity remained low in the relatively unfavorable summer of 1966.

Animals born in 1964, adult in 1967, also a large age-class, have a moderate start and go

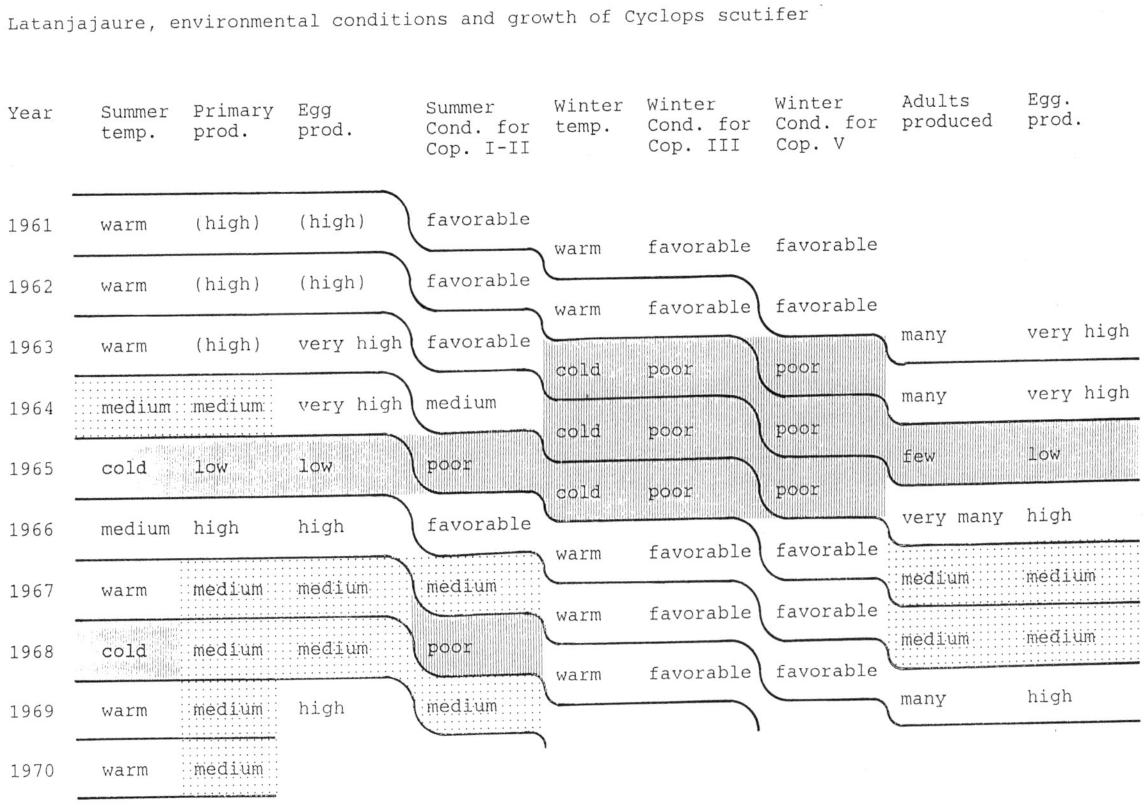

Fig. 11. Latnjajaure. Cyclops scutifer growth and environmental conditions 1961–1970.

through a cold winter as copepodide III, but a warm one as copepodide V. Losses again, were obvious; reproduction is good in the favorable summer of 1967 but because of low individual numbers, total egg production remains moderate. Animals born in 1965, adult in 1968, are a weak age-class and have an unfavorable start. Also as copepodides III they experience a poor summer. The situation improves for them as copepodides V, but the final result is mediocre. The last age-class whose life-time is covered by this study is born in 1966, adult in 1969. With moderate nauplii numbers at the beginning, this age-class spends all its life under reasonable or very good conditions and also ends up, after having suffered few losses, as a strong population with high individual and total reproduction.

Predation and competition

Predation on the *Cyclops scutifer population*, if significant at all, is due to cannibalism of the elder developing stages on the younger ones. However, the losses at different stages according to what is shown in Fig. 6 cannot clearly be bound to cannibalic feeding. A few *Megacyclops gigas* close to the bottom and a few *Dytiscus* larvae, mainly close to the shore, may catch one or the other *Cyclops*. One probably more important, interesting predator are the larvae of *Pseudodiamesa nivosa* GOETHG. This species in summer carries out diurnal migrations, up to the surface at night, where it is fixed by the surface tension. It seems that it adds a substance to the water which increases the surface tension so that migrating *Cyclops*, mainly the males, are caught in it. Thanks to this trick, the chironomid larvae which are poor swimmers, are enabled to feed on *Cyclops* of which their guts could be found completely stuffed.

There is no answer, at the moment, to the question of the quantitative importance of other zooplankters as food or as competitors of *Cyclops scutifer*. Rotifers may be eaten, but there are no signs that *Eubosmina* serves as a prey. *Eubosmina*, on the other hand, certainly feeds on small particles, mainly phytoplankton, and in this respect is a potential competitor of *Cyclops*. Different depth preferences and different horizontal distribution patterns (Nauwerck, 1978) may be seen as competition avoidance strategies.

Discussion

Cyclops scutifer is a widespread species of the northern subarctic belt and has been studied in many lakes between Kamtchatka, northern America and Scandinavia. Remarkable variability has been stated concerning the species' life cycle (listed e.g. by Paquette & Pinel-Alloul, 1982). There are one year cycles without diapause, mainly in moderately cool lakes, two and three year cycles without diapause in cold and very cold lakes, there are one to three year cycles, possibly even four year cycles, interrupted by diapause, mainly in tarns, often such with hypolimnic oxygen depletion, and further complications are caused by the separation of fractions wherof one may develop in a one year cycle, the other, with or without diapause, in a two year cycle or more. Elgmork *et al.* (1978) and Nilssen (1978) suggest adaptation to biological environmental pressure (predation at crucial times of development, even competition) as an explanation. Yet knowledge is limited as to how far this wide range of life strategies is a result of ecological flexibility and how far it reflects established genetical differences.

The primary problem that *Cyclops scutifer* faces in an arctic-alpine lake like Latnjajaure is the problem of very low temperatures. During 10 months of winter the water keeps 1–2 °C, maybe one more degree close to the bottom. During a few summer months water temperature may increase to 6–8 °C. The *Cyclops* solves this problem simply by slow development. This, on the other hand, enables it to make better use of the limited food resources, and to store, during a long existence at juvenile stages, enough reserves to reproduce independently of the food supply it meets when it finally moults to adult. Nevertheless, actual summer food supply may result in further egg production during the season. Absence of fish and low predation pressure allows population survival in spite of low reproduction and very long exposition, during juvenile life, to environmental injuries.

Light obviously plays a role when the *Cyclops* population starts moving up when the lake is still covered with ice. Another trigger may be the melting water which mixes into the lake already before ice-break importing organic and inorganic substances from the surroundings. During the time of mating and reproduction, the population shows a considerable oscillation in time and space, which may be an advantage in partner search and genetical mixing. As not only zooplankton distribution but also phytoplankton distribution in the lake is very patchy (Nauwerck, 1980; Isaksson, unpublished), this mobility may also include food searching advantages.

In other cases where horizontal and vertical distribution have been studied, pronounced patchiness and preference of deeper water layers by the copepodides, particularly during winter were also found. Diurnal vertical migration has been demonstrated from several lakes. Largest migration amplitudes appear in summer and in the more advanced stages, while young copepodes and nauplii show little migration or no migration at all (Nosova, 1972; Halvorsen & Elgmork, 1976; Larson, 1978; Elgmork & Eie, 1989). In contrast to normal night upwards, day downwards movement, Cunningham (1972) describes the opposite from an ice covered lake. Similar observations on temporal changes of migration behaviour were made on *Eudiaptomus gracilis* in Lake Erken (Nauwerck, 1963). It has to be pointed out that migration in Latnjajaure occurs in spite of the lack of fish. Predation pressure therefore must be ruled out as a reason for migration.

There is a general agreement between different authors' observations on the food of *Cyclops scutifer* and actual findings. *Cyclops scutifer* may feed on phytoplankton mainly (Nosova, 1972; Persson, 1985), or even take zooplankton, particularly rotifers and nauplii (Strickler & Twombly, 1972; Boers & Carter, 1978; Moore 1978). Even ciliates may play a role in its nutrition (Porter *et al.*, 1979). Besides fish and *Chaoborus*, *Megacyclops gigas* has been mentioned as a possible predator (Larson, 1978; Nilssen, 1978). New is, in our case, the controlling role of chironomid larvae.

References

Axelson, J., 1961. On the dimorphism in *Cyclops scutifer* (SARS) and the cyclomorphosis in *Daphnia galeata* (SARS). Rep. Inst. Freshwat. Res. Drottningholm 42: 169–181.

Bodin, K., 1966. Produktionsbiologiska studier över *Marsupella aquatica* (SCHRADER) SCHIFFNER i Latnjajaure. Scripta limnol. upsal., Coll. 2, Limnol. Inst. Uppsala 43 pp.

Bodin, K. & A. Nauwerck, 1968. Produktionsbiologische Studien über die Moosvegetation eines klaren Gebirgssees. Schweiz. Z. Hydrobiol. 30: 318–352.

Boers, J. J & J. C. H. Carter, 1978. The life history of *Cyclops scutifer* SARS (Copepoda, Cyclopoida) in a small lake of the Matamek River System, Quebec. Can. J. Zool. 56: 2603–2607.

Cunningham, L., 1972. Vertical migrations of *Daphnia* and copepods under the ice. Limnol. Oceanogr. 17: 301–303.

Ekman, S., 1904. Die Phyllopoden, Cladoceren und freilebenden Copepoden der nordschwedischen Hochgebirge. Zool. Jahrbücher, 21, Abt, f. System.

Ekman, S., 1957. Die Gewässer des Abisko-Gebietes und ihre Bedingungen. Kungl. Svenska Vetenskapsakad. Handl., Ser. IV, Bd. 6, Nr. 6, 172 pp.

Elgmork, K., 1965. A triennial copepod in the temperate zone. Nature 205: 413.

Elgmork, K., 1981. Extraordinary prolongation of the life cycle in a freshwater planktonic copepod. Holarct. Ecol. 4: 278–290.

Elgmork, K. & A. Langeland, 1980. *Cyclops scutifer* 1 and 2 year life cycles with diapause in the meromictic lake Blankvatn, Norway. Arch. Hydrobiol. 88: 178–201.

Elgmork, K. & J. A. Eie, 1989. Two- and three-year life cycles in the planktonic copepod *Cyclops scutifer* in two high mountain lakes. Holarctic. Ecol. 12: 60–69.

Elgmork, K., J. P. Nilssen, T. Broch & R. Øvrevik, 1978. Life cycle strategies in neighbouring populations of the copepod *Cyclops scutifer* SARS. Verh. int. Ver. Limnol. 20: 2518–2523.

Elgmork, K., G. Halvorsen, J. A. Eie & A. Langeland, 1990. Coexistance with similar life cycles in two species of freshwater copepods, Crustacea. Hydrobiologia 208: 187–200.

Ericsson, G. & G. Persson, 1971. Limnologiska studier i högfjällsvattenn i Latnjajaureområdet 1967. Scr. limnol. upsal., Collect. 7A, ser. 278, 79 pp.

Halvorsen, G. & K. Elgmork, 1976. Vertical distribution and seasonal cycle of *Cyclops scutifer* SARS (Crustacea, Copepoda) in two oligotrophic lakes in southern Norway. Norw. J. Zool. 24: 143–160.

Hellström, B. G. & A. Nauwerck, 1971. Zur Biologie und Populationsdynamik von *Polyartemia forcipata* (FISHER). Rep. Inst. Freshwat. Res. Drottningholm 51: 47–66.

Larsson, P., 1978. The life cycle dynamics and production of zooplankton in Øvre Heimdalsvatn. Holarct. Ecol. 1: 162–218.

Lindström, T., 1958. Observations sur les cycles annuels des planctons crustacés. Rep. Inst. Freshwat. Res. Drottningholm 39: 99–145.

Lithner, G., 1966. Bottenfaunan i Latnjajaure. Stenzil, Limnol. Inst. Uppsala, 33 pp.

Lohammar, G., 1963. Om Torneträsk och kringliggande sjöar. Natur i Lappland. K. Curry-Lindahl ed., Bokförlaget Svensk Natur, Upsala.

McLaren, I. A., 1961. A biennial copepod from Lake Hazen, Ellesmere Island. Nature 189: 774.

Moore, J. W., 1978. Composition and structure of zooplankton communities in 18 arctic and subarctic lakes. Int. Revue Hydrobiol. 63: 545–565.

Nauwerck, A., 1963. Die Beziehungen zwischen Zooplankton und Phytoplankton im See Erken. Symb. Bot. Upsal. XVII: 5, 163 pp.

Nauwerck, A., 1966. Beobachtungen über das Phytoplankton klarer Hochgebirgsseen. Schweiz. Z. Hydrol. 28: 4–28.

Nauwerck, A., 1967. Das Latnjajaureprojekt. Untersuchung eines fischfreien Sees vor und nach dem Einsatz von Fisch. Rep. Inst. Freshwat. Res. Drottningholm 47: 56–75.

Nauwerck, A., 1968a. Das Phytoplankton des Latnjajaure 1954–1965. Schweiz. Z. Hydrol. 30: 188–216.

Nauwerck, A., 1968b. Bottenfaunan i Latnjajaure 1964 enligt Jan Karlssons preliminära undersökningar. Stenzil, Limnol. Inst. Uppsala, 14 pp.

Nauwerck, A., 1978. *Bosmina obtusirostris* SARS im Latnjajaure. Arch. Hydrobiol. 82: 387–418.

Nauwerck, A., 1979. Zur Gattung *Chrysolykos* MACK. Bot. Notiser 132: 161–183.

Nauwerck, A., 1980. Die pelagische Primärproduktion im Latnjajaure, Schwedisch Lappland. Arch. Hydrobiol./ Suppl. 57: 291–323.

Nauwerck, A., 1981. Studien über die Bodenfauna des Latnjajaure (Schwedisch Lappland). Ber. nat.-med. Ver. Innsbruck 68: 79–98.

Nauwerck, A., 1983. Vattenkemi och plankton i sjöar i Abiskoområdet augusti 1981. Länsstyrelsens i Norrbottens läns rapportserie 1983: 14, 26 pp.

Nauwerck, A. & G. Persson, 1971. Ekmanjaure – en återfödd sjö. Fauna och Flora 66: 130–140.

Nilssen, J. P., 1978. On the evolution of life histories of limnetic cyclopoid copepods. Mem. Ist. ital. Idrobiol. 36: 193–214.

Nosova, I. A., 1972. Diurnal migrations of *Cyclops scutifer* (Copepoda, Cyclopoida) in the Kuril Lake (Kamtchatka). Zool. Zh. 51: 1457–1476.

Palm, B. & S. Törmä, 1965. Latnjavaggejaure. Geomorfologiska studier i Latnjavagge och morfometrisk undersökning av Latnjavaggejaure. Stenzil, Geogr. Inst. Uppsala, 28 pp.

Paquette, M. & B. Pinel-Alloul, 1982. Developmental cycles of *Skistodiaptomus oregonensis*, *Tropocyclops prasinus* and *Cyclops scutifer* in the limnetic zone of Lake Chromwell, St. Hippolyte, Quebec, Canada. Can. J. Zool. 60: 139–151.

Pejler, B., 1957. Taxonomical and ecological studies on planktonic Rotatoria from northern Swedish Lapland. K. Svenska VetenskAkad. Handl., ser. 4, Bd. 6, nr. 5, 68 pp.

Persson, G., 1985. Community grazing and the regulation of *in situ* clearance and feeding rates of planktonic crustaceans in lakes of the Kuokkel Area, Northern Sweden. Arch. Hydrobiol. Suppl. 70: 197–238.

Porter, K. G., M. L. Pace & J. F. Battey, 1979. Ciliate protozoans as links in fresh water planktonic food chains. Nature 277: 563–565.

Rodhe, W., 1962. Sulla produzione di fitoplankton in laghi trasparenti di alta montagna. Mem. Ist. ital. Idrobiol. 15: 21–28.

Rodhe, W., 1963. Livet i högfjällssjöarna. Natur i Lappland. K. Curry-Lindahl ed., Bokförlaget Svensk Natur, Uppsala: 184–189.

Rodhe, W., J. E. Hobbie & R. T. Wright, 1966. Phototrophy and heterotrophy in high mountain lakes. Verh. int. Ver. Limnol. 16: 302–313.

Schönborn, W., 1973. Paläolimnologische Studien aus Bohrkernen des Latnjajaure (Abisko-Gebiet; Schwedisch-Lappland). Hydrobiologia 42: 63–75.

Schönborn, W., 1975. Studien über die Testaceenbesiedlung der Seen und Tümpel des Abisko-Gebietes (Schwedisch-Lappland). Hydrobiologia 46: 115–139.

Silina, N. I., 1988. Features of the biology of *Cyclops scutifer* Sars, Copepoda, Cyclopoida in Lakes in the Central Yakut ASSR, Russian SFSR, USSR. Gidrobiol. Zh. 24: 78–79.

Strickler, J. R. & S. Twombly,1975. Reynolds number, diapause and predatory copepods. Verh. int. Ver. Limnol. 19: 2951–2958.

Taube, I., 1966. Embryonalutvecklingens beroende av temperaturen hos *Mesocyclops leuckarti* (CLAUS) och *Cyclops scutifer* SARS. Stenzil, Limnol. Inst. Uppsala, 18 pp.

Taube, I. & A. Nauwerck, 1967. Zur Populationsdynamik von *Cyclops scutifer* SARS. I. Die Temperaturabhängigkeit der Embryonalentwicklung von *Cyclops scutifer* SARS im Vergleich zu *Mesocyclops leuckarti* (CLAUS. Rep. Inst. Freshwat. Res. Drottningholm 47: 76–86.

Wåhlin, I., 1970. Die Diatomeen des Latnjajaure. I. Die rezenten Bodendiatomeen. Arch. Hydrobiol. 67: 460–484.

Hydrobiologia **274**: 115–120, 1994.
J. Fott (ed.), Limnology of Mountain Lakes.

Comparison of diatom communities in remote high-mountain lakes using index B and cluster analysis

Pius Niederhauser & Ferdinand Schanz
University of Zürich, Institute of Plant Biology, Limnological Station, Seestrasse 187, CH-8802 Kilchberg, Switzerland

Key words: diatoms, pH, index B, high-mountain lakes, cluster analysis

Abstract

Six remote high-mountain lakes in SE Switzerland were characterized on the basis of their water chemistry and the composition of their epilithic diatom communities. The index B (Renberg & Hellberg, 1982), calculated by summarizing the relative frequencies of occurrence in all five pH classes defined by Hustedt (1938/39), was found to be well correlated with pH values measured from August to October (5.8 to 7.0). Four lakes with pH values from 6.3 to 6.6 could not be distinguished from one another by means of the index B despite some differences in their physical, chemical and biological characteristics. However, it was possible to separate them by cluster analysis using the relative frequencies of the diatom species. We conclude that much ecological information is lost when applying the pH classification of Hustedt (1938/39).

Introduction

A decrease in pH has a pronounced effect on the diatom communities of freshwater ecosystems (Charles *et al.*, 1989). Diatom frustules in sediments are generally well preserved, allowing the reconstruction of the pH history of a lake based on the stratigraphy of subfossil diatom assemblages (Nygaard, 1956). The methods of Nygaard (1956), Meriläinen (1967), Renberg & Hellberg (1982), Davis & Anderson (1985) and Charles (1985), which were developed to infer the pH of lakes from diatom assemblages, are all based on the pH classification system of Hustedt (1938/39). According to calibration data based on recent diatom assemblages (Charles, 1985; Jones & Flower, 1986; Arzet, 1987) index B of Renberg & Hellberg (1982) indicates a good accordance between measured and inferred pH in the range of pH 5.5 to 7.0. The intentions of our study was to find out the application limits of index B when compared the epilithic diatom communities of six high-mountain lakes with chemical and physical properties of the lake waters. The classification based on index B was checked using cluster analysis.

Area of investigation

The Macun lakes are situated in SE Switzerland near the Swiss National Park at about 2600 m above sea level (Fig. 1). Morphometric data are summarized in Table 1. The drainage area is not populated and consists of crystalline rocks, mainly of gneiss. The direct influence of human and animal activities on lakes A, B, C, D and E is negligible. Near lake M, however, a military refuge-hut is occupied periodically. It is possible that nutrients and neutralizing substances which

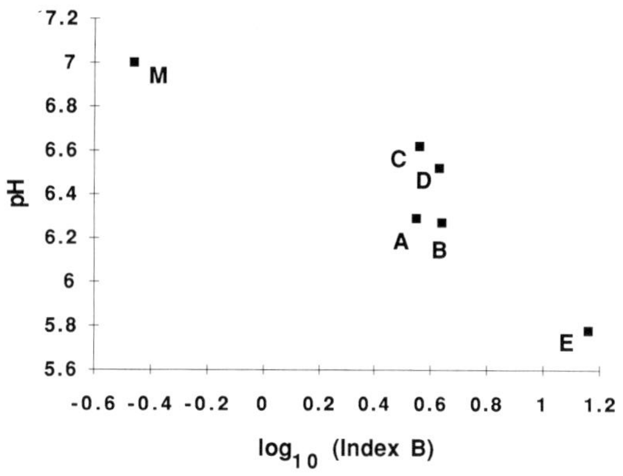

Fig. 3. Correlation between the logarithms ($\log_{10}$) of index B (after Renberg & Hellberg, 1982) and the measured mean summer pH values of the Macun lakes A to E and M. Calculation:Index B = ((%ind) + 5·(%acf) + 40·(%acb)) / ((%ind) + 3.5·(%alkf) + 108·(%alkb)); Acb = acidobiontic, acf = acidophilic, ind = indifferent, alkf = alkaliphilic, alkb = alkalibiontic.

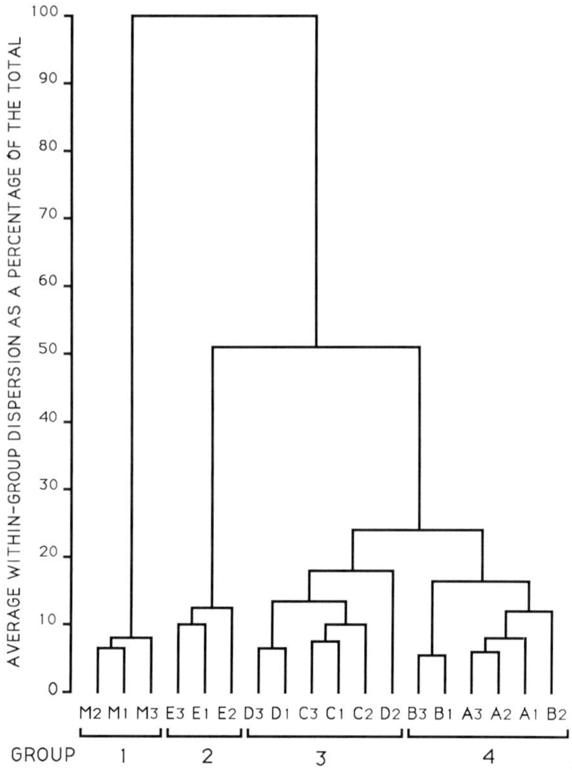

Fig. 4. Classification of epilithic diatom communities in the Macun lakes using cluster analysis. The species included in the calculations are those for which the relative frequency of occurrence exceeded 1% of the total number of diatom frustules in at least one sample. The analysis is based on three samplings from each of the lakes (lakes A to E, August to October 1987; lake M, August 1988).

of the periphyton diatom community of lake M deviates from that of all other lakes (dominance of *Cymbella minuta* Hilse, *Fragilaria pinnata* und *Nitzschia perminuta* (Grun.) Peragallo).

The degree of correlation between the calculated indices B and the measured pH values is apparent from Fig. 3. The distances between lake E, lake M and the cluster formed by lakes A to D are considerable. However, it is evident that lakes A to D cannot be separated using log index B.

The cluster analysis based on the data of the epilithic diatom communities (Fig. 4) results in a classification slightly different from that based on log index B. As in Fig. 3, lake E (group 1) and lake M (group 2) are separated from the other lakes; but additionally, lakes C and D (group 3) and lakes A and B (group 4) form two distinct clusters. Furthermore, a high degree of similarity is obvious between the diatom communities taken from lakes A and B, as well as between those taken from lakes C and D.

Discussion

Based on the low conductivity values (3.4 to 8.3 $\mu S\ cm^{-1}$ Table 2) we would expect lakes A to E to be sensitive to acidification processes (Mosello, 1983; Psenner *et al.*, 1988). However, average pH values during the summer months have ranged between 5.8 and 6.6 during the last few years (Schanz, 1987), and so there is no evidence of acidification so far. Diatom analyses from sediment cores confirm this (Niederhauser, unpubl. data). Index B values computed from the core sections did not increase towards the sediment surface. We conclude that the anthropogenic input of acids by precipitation is of minor importance in the Macun area.

Lake M has a conductivity of 39 $\mu S\ cm^{-1}$ and therefore has higher electrolyte concentrations than the other Macun lakes, possibly because of the nearby military refuge. It is therefore not surprising that the diatom community of lake M is distinctly different from diatom communities of the other lakes (Fig. 2).

The species *Achnanthes helvetica*, *A. grischuna*, *A. marginulata* and *Tabellaria flocculosa* are acidophilic forms (preferred pH$\leq$7). These species account for 37% of all diatoms in lake A, 43% in lake B, 32% in lake C; 28% in lake D and 52% in lake E. These frequencies correspond well with the pH values presented in Table 2. Lakes C and D have the highest pH values (6.6 and 6.5), lakes A and B have a slightly lower pH (both 6.3) and the lowest pH was found in lake E (5.8). However, the alkalophilic species *Fragilaria pinnata* occurs more frequently in lakes A (3%) and B (2%) than in lakes C and D (both<0.5%), disagreeing with the result which would be expected based on the measured pH. The species *Fragilaria pinnata* occurs most frequently in lake M (29%).

Using cluster analysis, lakes A and B and lakes C and D were separated into two clusters (Fig. 3). Differences in the diatom community structures of the two clusters are partly recognizable in Fig. 2 (e.g. *Achnanthes helvetica, A. grischuna*). However, these differences are not apparent in the index B statistics. We presume that the separation of the four lakes A, B, C and D into two groups is an effect of including less frequent species (not illustrated in Fig. 2) in the cluster analysis. It seems in general that multivariate statistical methods are well suited to deal with the multidimensional data sets involved in the analysis of diatom community structures (Huttunen & Meriläinen, 1986). Multivariate methods are based on much more information then are index B statistics, leading to a better interpretation of the results with regard to their ecological significance.

Renberg & Hellberg (1982) found a standard deviation of 0.3 pH units for the regression equation of index B, which is in good correspondence with the results of other authors (Charles, 1985; Arzet, 1987). This means that lakes with pH values differing by less than 0.3 pH units cannot be expected to have significant differences in their index B values. This was the case in lakes A to D, where pH values ranged between 6.3 and 6.6, and could explain the inability of the index B method to distinguish between these lakes.

Ter Braak & Van Dam (1989) found that the method of weighted averaging gives better results for the characterization of the pH of lakes than do index B statistics. However, they needed detailed descriptions of the pH optima of the diatom species occurring which are not yet available for the alpine region. This method could not therefore be used in our study. As with multivariate statistical analysis, the employment of weighted averaging would increase the ecological significance of the results by considering a more or less complete set of information about the diatom communities found. This contrasts with the index B calculations, where all the information is reduced to five pH classes.

Conclusions

In general the index B was well correlated with the pH values measured. However, those Macun lakes with pH values differing by less than 0.3 units could not be separated from one another. Despite this, the structure of their diatom communities and their chemical properties differed considerably.

A calibration set for the alpine region is required giving detailed descriptions of the pH optima of the diatom species occurring.

Multivariate methods, such as cluster analysis, are well suited to deal with the multidimensional data sets involved in the analysis of diatom community structures.

Acknowledgements

We are indebted to Dr J. Hürlimann for his help in determining the diatom species, to H. P. Mächler for the chemical analysis and to Dr D. M. Livingstone for correcting the English text. The field work was supported by the 'Kommission für die wissenschaftliche Erforschung des Nationalparks' of the Swiss Academy of Sciences and also by the Swiss National Science Foundation (Grant 31-27375.89).

References

Anderson, D. S., R. B. Davis & F. Berge, 1986. Relationship between diatom assemblages in lake surface-sediments and limnological characteristics in southern Norway. In J. P. Smol, R. W. Batterbee, R. B. Davis & J. Meriläinen (eds), Diatoms and lake acidity. Dev. Hydrobiol. 29. Dr W. Junk Publishers, Dordrecht: 97–114.

Arzet, K., 1987. Diatomeen als pH-Indikatoren in subrezenten Sedimenten von Weichwasserseen. Diss. Abt. Limnol. Innsbruck 24: 1–266.

Camburn, K. E. & J. C. Kingston, 1986. The genus Melosira from softwater lakes with special reference to northern Michigan, Wisconsin and Minnesota. In J. P. Smol, R. W. Batterbee, R. B. Davis & J. Meriläinen (eds), Diatoms and Lake Acidity. Dev. Hydrobiol. 29. Dr W. Junk Publishers, Dordrecht: 17–34.

Charles, D. F., 1985. Relationships between surface sediment diatom assemblages and lakewater characteristics in Adirondack lakes. Ecology 66: 994–1011.

Charles, D. F., R. W. Batterbee, I. Renberg, H. van Dam & J. P. Smol, 1989. Paleoecological analysis of lake acidification trends in North America and Europe using diatoms and chrysophytes. In S. A. Norton, S. E. Lindberg & S. L. Page (eds), Acidic precipitation. Soil, aquatic processes, and lake acidification, Vol. 4. Springer, N.Y., 293 pp.

Cholnoky, B. J., 1968. Die Oekologie der Diatomeen in Binnengewässern. J. Cramer Verlag, Lehre, 699 pp.

Davis, R. B. & D. S. Anderson, 1985. Methods of pH calibration of sedimentary diatom remains for reconstructing history of pH in Lakes. Hydrobiologia 29: 73–86.

Germain, H., 1981. Flore des Diatomées. Société nouvelle des éditions Boubée, Paris, 444 pp.

Hustedt, F., 1930. Bacillariophyta (Diatomeae). In A. Pascher (ed.), Süsswasser-Flora Mittel-Europas, Band 10. Fischer, Jena, 462 pp.

Hustedt, F., 1938/39. Systematische und ökologische Untersuchungen über die Diatomeenflora von Java, Bali und Sumatra. Arch. Hydrobiol., Suppl. 15: 638–790 & 16: 274–394.

Huttunen, K., & J. Meriläinen, 1986. Applications of multivariate techniques to infer limnological conditions from diatom assemblages. In J. P. Smol, R. W. Batterbee, R. B. Davis & J. Meriläinen (eds), Diatoms and Lake Acidity. Dev. Hydrobiol. 29. Dr W. Junk Publishers, Dordrecht: 201–211.

Jensen, S., 1980. Influences of transformation of cover values on classification and ordination of lake vegetation. Vegetatio 37: 19–31.

Jones, V. J. & R. J. Flower, 1986. Spacial and temporal variability in diatom communities: Palaeoecological significance in an acidified lake. In J. P. Smol, R. W. Batterbee, R. B. Davis & J. Meriläinen (eds), Diatoms and Lake Acidity. Dev. Hydrobiol. 29. Dr W. Junk Publishers, Dordrecht: 87–94.

Krammer, K. & H. Lange-Bertalot, 1986. Bacillariophyceae 1. Teil: Naviculaceae. In H. Ettl, J. Gerloff, H. Heynig & D. Mollenhauer (eds), Süsswasserflora von Mitteleuropa, Band 2/1, Fischer Verlag, Stuttgart, 876 pp.

Krammer, K. & H. Lange-Bertalot, 1988. Bacillariophyceae 2. Teil: Bacillariophyceae, Epithemiaceae, Surirellaceae. In H. Ettl, J. Gerloff, H. Heynig & D. Mollenhauer (eds), Süsswasserflora von Mitteleuropa, Band 2/1, Fischer Verlag, Stuttgart, 596 pp.

Meriläinen, J., 1967. The diatom flora and hydrogen-ion concentration of the water. Ann. bot. fenn. 4: 51–58.

Mosello, R., 1983. Hydrochemistry of high altitude lakes. Schweiz. Z. Hydrol. 46: 86–99.

Nygaard, G., 1956. Ancient and recent flora of diatom and chrysophyceae in lake Gribsø. Folia limnol. scand. 8: 32–99.

Psenner, R., U. Nickus & F. Zapf, 1988. Versauerung von Hochgebirgsseen in kristallinen Einzugsgebieten Tirols und Kärntens. Bundesministerium für Land- und Forstwirtschaft, Vienna, 335 pp.

Renberg, I. & T. Hellberg, 1982. The pH history of lakes in southwest Sweden, as calculated from the subfossil diatom flora of the sediments. Ambio 11: 30–33.

Schanz, F., 1984. Chemical and algological characteristics of five high mountain lakes near the Swiss National Park. Verh. int. Ver. Limnol. 22: 1066–1070.

Schanz, F., 1987. Beurteilung des Einflusses von sauren Niederschlägen auf das Macun-Gebiet im Unterengadin (Schweiz). Verh. Ges. Oekol. (Graz 1985) 15: 249–255.

Straub, F., 1981. Utilisation des membranes filtrantes en téflon dans la préparation des diatomées épilithique. Comptes rendus du 2e colloque de l'ADLAF. Cryptogamie. Algologie 2: 153.

Stumm, W. & J. J. Morgen, 1981. Aquatic chemistry. 2nd edn. Wiley Sons, N.Y., 780 pp.

Ter Braak, C. J. F. & H. van Dam, 1989. Inferring pH from diatoms: a comparison of old and new calibration methods. Hydrobiologia 178: 209–223.

Wildi, O. & L. Orlóci, 1990. Numerical exploration of community patterns. SPB Academic Publishing, The Hague, 124 pp.

Hydrobiologia **274**: 121–126, 1994.
J. Fott (ed.), Limnology of Mountain Lakes.

Zooplankton decline in the Černé Lake (Šumava Mountains, Bohemia) as reflected in the stratification of cladoceran remains in the sediment

Miroslava Pražáková & Jan Fott
Department of Hydrobiology, Faculty of Science, Charles University, Viničná 7, 128 44 Prague 2, Czech Republic

Key words: Cladocera remains, lake sediment, palaeolimnology

Abstract

Stratigraphy of cladoceran remains in the upper 18 cm of a sediment core from the Lake Černé was studied. The successive disappearance of *Bosmina longispina*, *Daphnia longispina* and *Ceriodaphnia quadrangula* from the upper layers of the sediment corresponds with our knowledge concerning the disappearance of these species from the open water.

Introduction

Lake Černé (A = 18.4 ha, Z_{max} = 39 m, Z_m = 15.6 m, altitude = 1008 m) is a small lake of glacial origin situated in the Šumava Mountains, Bohemia. Its zooplankton have been studied since 1871. Although the data are scarce, there is clear evidence of a gradual decline in the six species of planktonic Crustacea present in the lake from 1871 until now, when there are none. The reasons of the decline are not fully understood, but the absence of planktonic Crustacea at present can be explained by acidification of the lake to pH 4.4–4.8 resulting in high concentrations of toxic aluminium ions and possibly other toxic substances as well (Fott *et al.*, 1993). The objective of this study is to compare the stratification of cladoceran remains in the upper sediment with our knowledge of the zooplankton decline during the last 100 years. Such a comparison may contribute to a better understanding of palaeolimnological evidence of succession processes in lakes.

Methods

A core 18 cm long was taken on September 27, 1979 from a 20 m depth, using a Jenkin sampler (Edmondson & Winberg, 1971). The core was cut into 18 segments of 1 cm height and analysed for remains of Cladocera. Each segment was diluted to about 50 ml by 10% KOH and stirred on a magnetic stirrer hot plate for half an hour. After digestion, the sample was passed through a metallic screen of 40 μm mesh size and washed with a large volume of distilled water. Mineral particles, pollen grains, plant fragments and cladoceran remains were retained on the screen, resuspended, and sedimented or centrifuged. The final volume was adjusted to 10 ml. A 1 ml subsample was transferred to glycerine-alcohol and stained by chlorazol black (Brandlová *et al.*, 1972). After the evaporation of water and alcohol, permanent mounts were made using polyvinyl alcohol. Cladoceran remains were determined according to Frey (1959) and counted at 100 × magnification. In addition, remains were compared with living specimens collected by the authors. All remains of every 1 ml subsample were counted. Head shields, shells, postabdomens, ephippia, and their fragments were counted, but only heads were used as an index of Chydoridae and Bosmina abundance. The head shields were more abundant than other remains of each species. Formulae for calculating the total numbers of remains were used according to 'Cladocera analysis' by Frey (1986). The results are expressed as numbers per 1 ml of the wet sediment.

Results

Data presented on the figures show the stratification of remains belonging to planktonic (*Bosmina longispina*, *Ceriodaphnia quadrangula* and *Daphnia longispina*) and littoral Cladocera (Chydoridae). The head shields of *Bosmina longispina* are absent in the uppermost top of the core. From the 4th cm downwards, their numbers increase. Ephippia of *Daphnia longispina* occur in the lower part of the core, and from the 9th cm upwards they are replaced by ephippia of *Ceriodaphnia quadrangula*. Postabdominal claws and postabdomens of *Daphnia* and *Ceriodaphnia* could not be distinguished and therefore were counted together. In the uppermost part of the core all remains of pelagic Cladocera are very scarce.

Remains of littoral Cladocera of the family Chydoridae were found in all layers of the core. *Alona affinis* is the most frequent species, forming a distinct peak between the 4th and 9th cm. *Acroperus harpae* and *Alonopsis elongata* were also found in the whole core, more frequently in the upper layers. Remains of *Alonella excisa* occurred mostly in the middle of the core. Other, less frequent species found in the core are *Alona quadrangularis*, *Alona guttata*, *Alona costata*, *Alonella exigua*, *Alonella nana*, *Chydorus sphaericus* and *Ch. piger*, *Eurycercus lamellatus*. Fig. 4 shows the number of all Chydoridae head shields found, with the highest frequency between the 4th and 10th cm.

Discussion

The possible reasons of the succession and eventual decline of planktonic Crustacea in the Lake Černé have been discussed elsewhere (Fott *et al.*, 1993). The main objective of the present study is comparison of the stratification pattern of cladoceran remains with the written information on the species composition and its changes in time. For this purpose the most suitable are remains of *Bosmina longispina*, *Daphnia longispina* and *Ceriodaphnia quadrangula*. They are well preserved in the sediment and their occurrence in the open water made the three species conspicuous to past investigators. Since 1871, opinions on their taxonomy and nomenclature developed (Table 1), but we do not suppose that more species of the genera *Bosmina*, *Daphnia* and *Ceriodaphnia* than the three quoted above were involved. The reports on their occurrence can be summarized as follows:

(1) *Bosmina longispina* was abundant in 1871 and 1892–96 (Frič, 1871; Hellich, 1877; Frič & Vávra, 1897). It has never been found since then.

(2) *Daphnia longispina* was abundant in 1871,

Table 1. Synonymics of three cladoceran species which had inhabited open water of the Černé Lake. The names used in the present study are in the headings of the columns. The taxonomic status of the species was determined after consultation with V. Kořínek.

	Bosmina longispina Leydig	*Daphnia longispina* O. F. Müller	*Ceriodaphnia quadrangula* (O. F. Müller)
Frič, 1871	*B. longispina*	*D. pulex et longispina*	
Hellich, 1877	*B. bohemica*	*D. ventricosa*	
Frič & Vávra, 1897	*B. bohemica*	*D. longispina ventricosa*	*C. pulchella*
Šrámek-Hušek, 1942	*B. longirostris bohemica*	*D. longispina ventricosa*	*C. quadrangula*
Šrámek-Hušek *et al.*, 1962	*B. coregoni longispina*	*D. longispina caudata*	*C. quadrangula*

Figs 1–3. Cladoceran remains in the upper 18 cm of the sediment. Numbers are expressed per ml of the wet sediment. Black columns: ephippia, white columns: postabdominal claws, dotted columns: head shields.

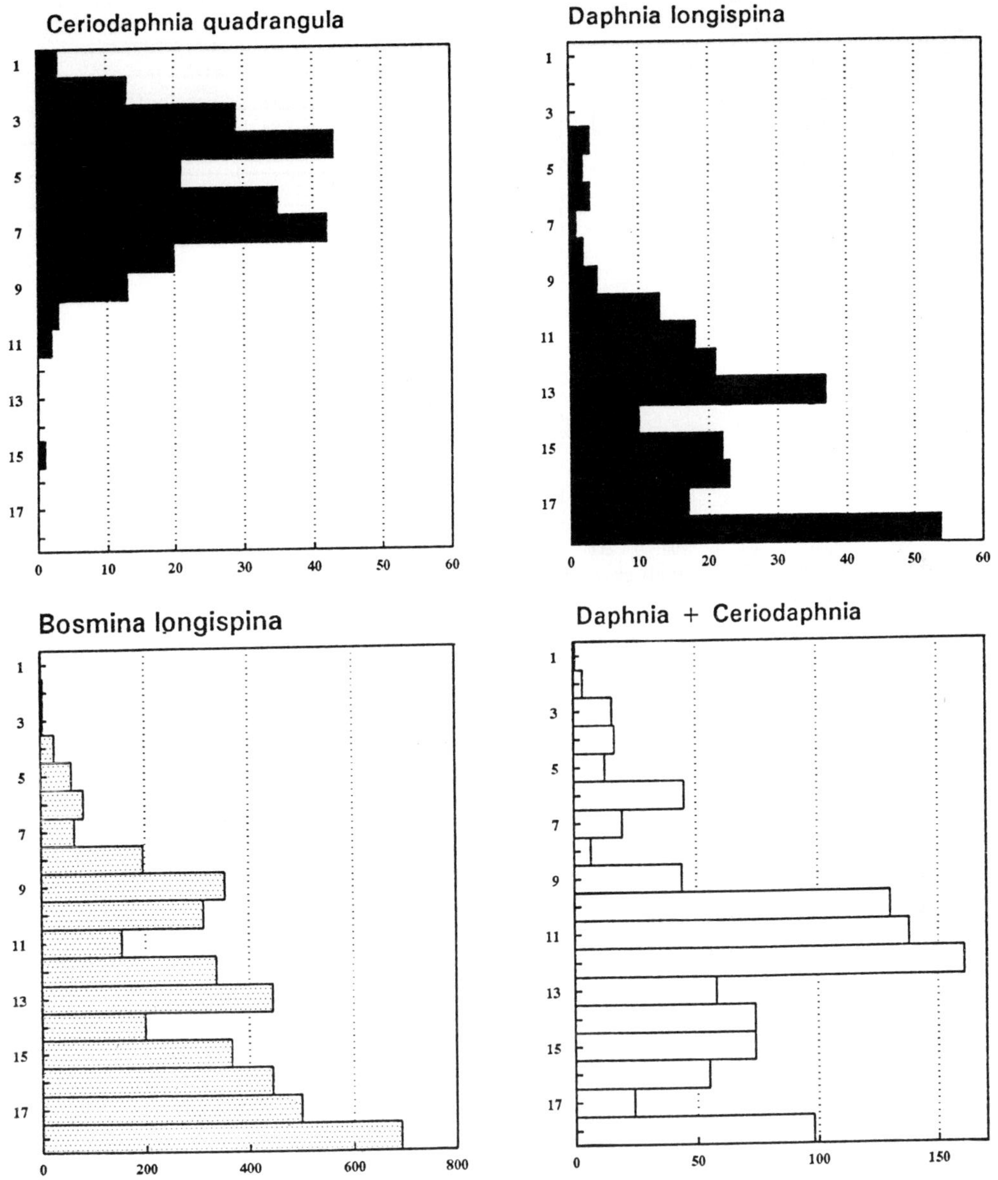

Fig. 1. Planktonic species.

but scarce in 1892–96 when it was sampled only once. (Frič, 1871; Hellich, 1877). Šrámek-Hušek (1942) found two specimens in 1935. No findings of this species have been reported later.

(3) *Ceriodaphnia quadrangula* was first mentioned from the period 1892–96 (Frič & Vávra, 1897), but only from the littoral. In 1935–37 it was the most abundant cladoceran, occurring in both the littoral and planktonic zones (Šrámek-Hušek, 1942). The species was reported also by other investigators until 1960; in 1979 we found one single specimen. The species has never been found since then.

The distribution patterns of *Bosmina*-head

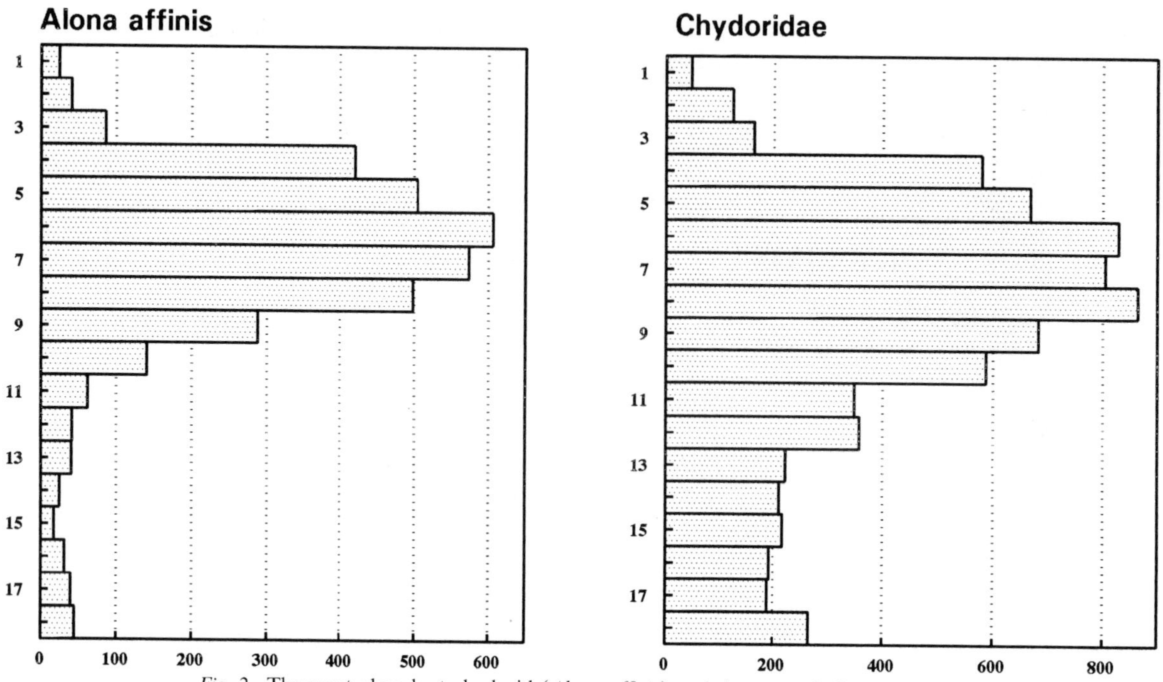

Fig. 2. The most abundant chydorid (*Alona affinis*) and the sum of all chydorids.

shields, *Daphnia*-ephippia, and *Ceriodaphnia*-ephippia are in accordance with the known history of decline of the three species (Figs 1, 2). The maximum abundance of *Ceriodaphnia*-ephippia is found in the upper layers where the *Bosmina*-head shields and *Daphnia*-ephippia are almost absent. It is difficult to decide whether the small numbers of *Bosmina* and *Ceriodaphnia* remains in the upper 7 cm resulted from an in-lake perturbation of the sediment, or whether they reflect a real presence of the populations not recorded by the earlier investigators.

The main factor controlling the abundance of remains of chydorid cladocerans in the sediment is the extent of the littoral zone (Harmworth & Whiteside, 1968), which may vary with the absorption of light in the water. At present, the littoral of the lake are stones and rock covered with a very thin, often almost invisible layer of periphyton. Higher plants were also scarce in the past (Frič & Vávra, 1987).

In a sample taken from shallow water above stones, in summer 1992, we found only 10 ind l^{-1} of *Alonopsis elongata*, and *Acroperus harpae* and *Alona guttata* in the densities of about 1 ind. l^{-1}. These low abundances of chydorid cladocerans correspond with low densities of their remains in the uppermost part of the sediment core. From the 4th cm downwards the abundances increase. There is an outstanding peak of *Alona affinis* between the 4th and 9th cm. Other chydorid species and the sum of chydorid remains have similar stratification patterns. The reasons of the recent decline seem to be the same as for the planktonic species: the severe acidification of the lake (more on that Fott *et al.*, 1994). The reasons of the increase in abundance of the chydorid remains from the 10th cm upwards is less clear. Perhaps it might indicate an increase in the extent and production of the epilithic algal zone, whatever the reasons.

The bottom of the Lake Černé (and also of the other lakes in Šumava) has been frequently disturbed by various human activities (J. Veselý, pers. comm.), which makes inference from sediment cores suspect. The different distinct patterns of cladoceran remains in the sediment core described here and the rough agreement of these patterns with the known history of the zooplankton of the lake, seem to prove that in this case we succeeded in taking a core of undisturbed sediment. If the patterns in repeatedly taken cores are consistent, then the subsampled cladoceran remains could be used for testing possible disturbance in cores taken for other, *i.e.* chemical, variables.

If the assumptions underlying the interpretation of the present data are correct, then the sediment accumulation rate in Lake Černé would be about 1 mm per year.

Acknowledgements

We would like to thank Vladimír Kořínek for help with the determination of cladoceran remains.

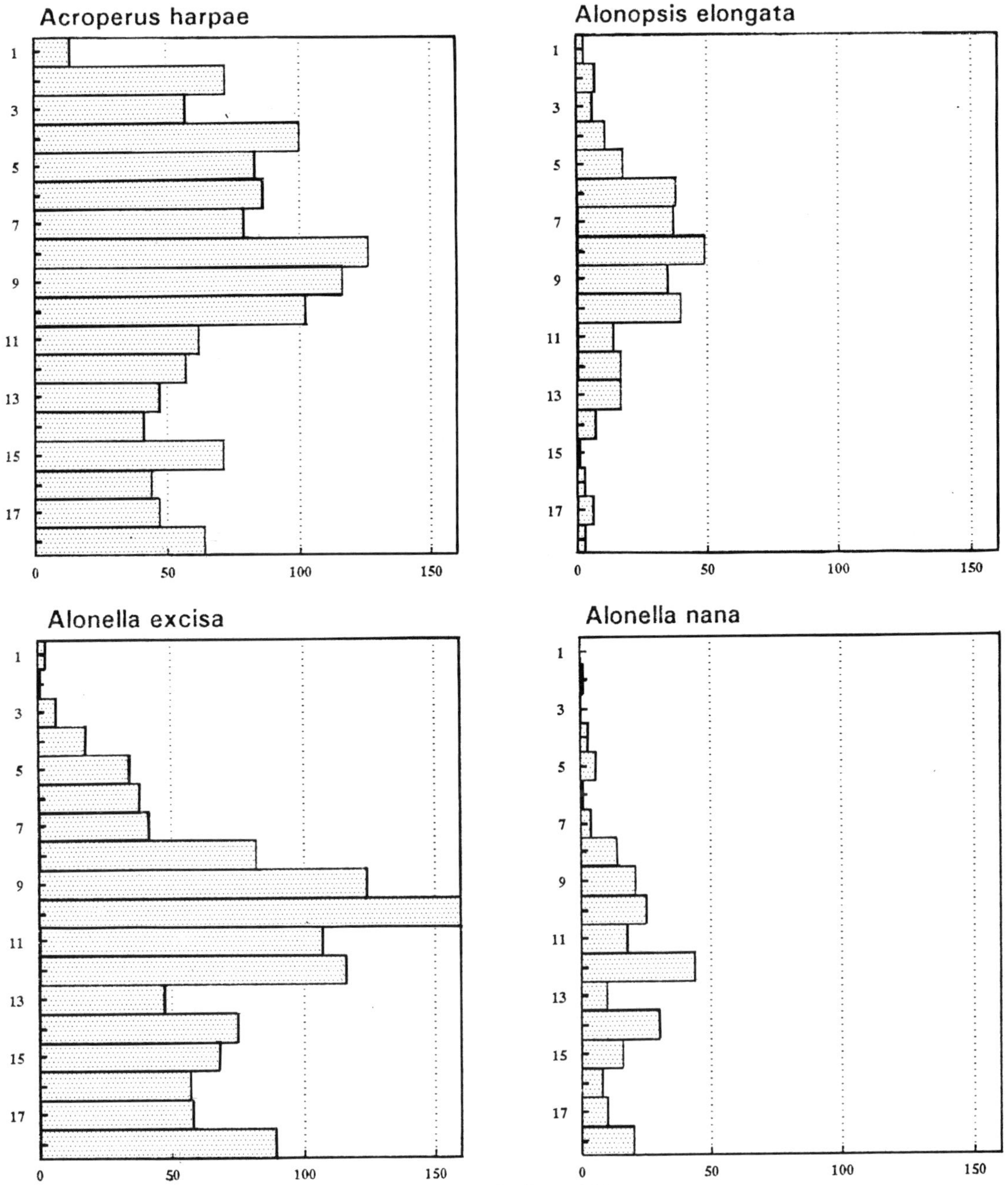

Fig. 3. See p. 126.

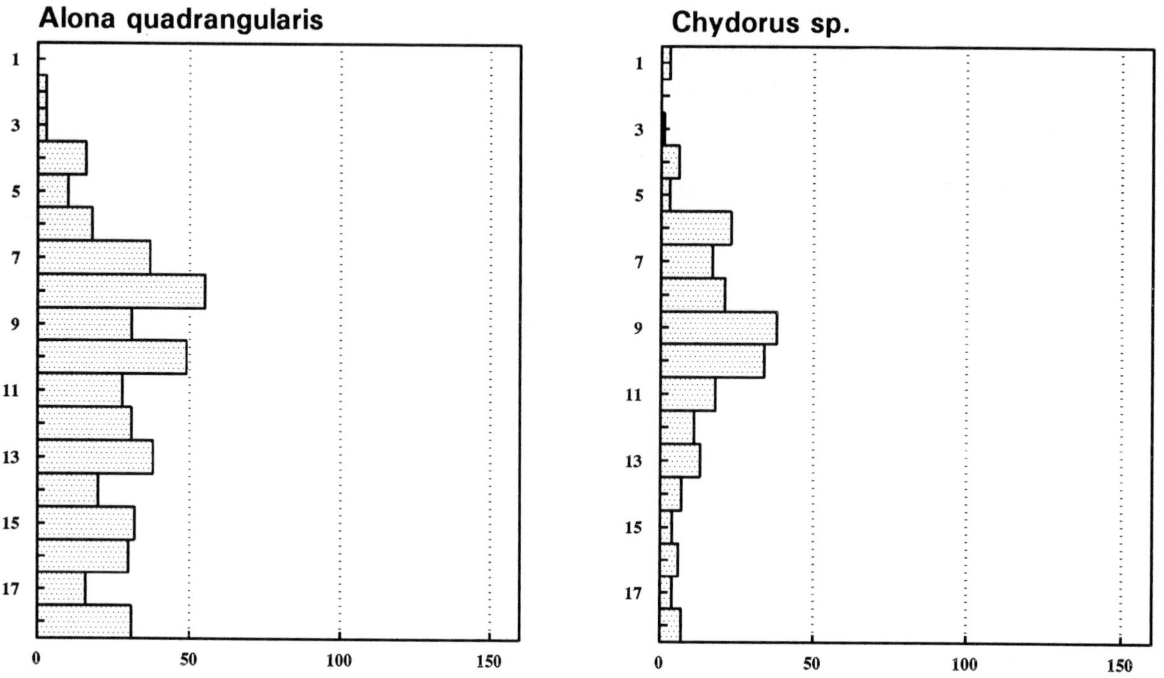

Fig. 3. Other chydorid species.

References

Brandlová, J., Z. Brandl & C. H. Fernando, 1972. The Cladocera of Ontario with remarks on same species and distribution. Can. J. Zool. 50: 1373–1403.

Edmondson, W. T. & G. G. Winberg, 1971. Secondary productivity in fresh waters. IBP Handbook No. 17, Blackwell Scientific Publications, Oxford, 358 pp.

Fott, J., E. Stuchlík & Z. Stuchlíková, 1987. Acidification of lakes in Czechoslovakia. In Moldan, B. & T. Pačes (eds), Extended abstracts of the International workshop on geochemistry and monitoring in representative basins (GEOMON). Geological Survey, Prague: 77–79.

Fott, J., M. Pražáková, E. Stuchlík & Z. Stuchlíková, 1994. Acidification of lakes in Šumava (Bohemia) and in the High Tatra Mountains (Slovakia). Hydrobiologia 274/Dev. Hydrobiol. 93: 37–47.

Frey, D. G., 1958. The late-glacial cladoceran fauna of a small lake. Arch. Hydrobiol. 54: 209–275.

Frey, D. G., 1959. The taxonomic and phylogenetic significance of the head pores of the Chydoridae (Cladocera). Int. Revue ges. Hydrobiol. 44: 27–50.

Frey, D. G., 1986. Cladocera analysis. In Berglund, B. E. (ed.), Handbook of Holocene Palaeoecology and Palaeohydrology. Wiley, Chichester: 667–692.

Frič, A., 1871. Über die Fauna der Böhmerwaldseen. Sitzungsber. d. k. böhm. Akad. d. Wiss. Prag, 13 pp.

Frič, A. & V. Vávra, 1897. Untersuchung zweier Böhmerwaldseen, des Schwarzen und des Teufelsees. Arch. d. naturw. Landesdurchforschung von Böhmen, 10, 74 pp.

Harmsworth, R. V. & M. C. Whiteside, 1968. Relation of Cladoceran remains in lake sediments to primary productivity of lakes. Ecology 49: 998–1000.

Hellich, B., 1877. Die Cladoceren Böhmens. Prague, 131pp.

Šrámek-Hušek, R., 1942. Revision der Cladoceren – Eucopepodenfauna des Schwarzen Sees in Böhmerwald nach 66 Jahren. (In Czech, with German summary). Mém. de la Soc. Royal de Boheme, Classe Sci., Prague, 22 pp.

Šrámek-Hušek, R., M. Straškraba & J. Brtek, 1962. Lupenonožci – Branchiopoda (in Czech). Fauna ČSSR 16, NČSAV, Prague, 470 pp.

Hydrobiologia **274**: 127–132, 1994.
J. Fott (ed.), Limnology of Mountain Lakes.

Characterization of carbonaceous particles from lake sediments

Neil Rose
Environmental Change Research Centre, Department of Geography, University College London, 26 Bedford Way, London WC1H OAP, UK

Abstract

Spheroidal carbonaceous particles produced by high temperature combustion of coal and oil are found in high concentrations in lake sediments from areas of high acid deposition. The sediment record of these particles showing the onset of industrialisation correlates well with the record of acidification as indicated by diatom analysis.

To find sources of the atmospheric deposition affecting a lake and its catchment, characterisation of the carbonaceous particles is necessary. A reference data set of particle chemistries from coal and oil power stations was produced using EDS generated data of 17 elements. Using multivariate statistical techniques, the most important elements for the coal/oil separation were identified and incorporated into a linear discriminant function which allocated fuel type with $>97\%$ accuracy.

Application of this technique to surface sediments in Scotland shows the influence of oil burning from outside the region, higher areas located on the east coast and in the south-west of the country. When applied to a full sediment core, the history of coal and oil combustion affecting the lake is seen and correlates well with known coal and oil consumption figures. Consequently this method could be used to add extra dating levels to sediment cores.

The technique has been extended to include peat particles and could potentially be used on those from brown coal, lignite and oil shale combustion.

Introduction

Lake sediments provide a record of atmospheric contamination and so have been important in recent studies of surface water acidification. Carbonaceous particles derived from fossil-fuel combustion are found in considerable numbers in upper levels of sediment cores taken from areas with high acid deposition (Griffin & Goldberg, 1981; Renberg & Wik, 1984). Sites in the United Kingdom show close correlation between the onset of atmospheric contamination as indicated by carbonaceous particles, heavy metals etc., and the acidification of lakes as indicated by diatom analysis (Battarbee *et al.*, 1988).

The particulate emissions from high temperature fossil fuel combustion can be divided into two groups, spheroidal carbonaceous particles, which are composed mainly of elemental carbon (Goldberg, 1985), and inorganic ash spheres, which are formed by the fusing of inorganic minerals within the fuel (Raask, 1984). Of the fossil fuels commonly used in Britain, only coal and oil produce spherical carbonaceous particles. Those produced from peat combustion have an amorphous appearance, many still retaining some cellular structure.

The inorganic ash spheres have not been characterised and this is because, being fused mineral inclusions their chemistry is independent of fuel type, and also, in Britain they are almost exclusively coal in origin.

Methods

The sediment cores were taken using a variety of methods (Livingstone, 1955; Kajak, 1966; Mackereth, 1969) and the carbonaceous particles extracted from the sediments as described in Rose (1990).

The analysis of the particle chemistries was undertaken at Imperial College using a JEOL 733 Superprobe linked to a Scanning Electron Microscope (SEM). These data were gathered using a fully automated technique as described in Watt (1990). Initial handling of these data used the MIDAS program before being transferred for further statistical analysis.

The Hampstead Heath core was dated in the Department of Applied Mathematics and Theoretical Physics, at the University of Liverpool using gamma spectrometry to analyse for ^{210}Pb, ^{226}Ra, ^{137}Cs and ^{241}Am (Appleby *et al.*, 1986).

A reference data set

In order to construct a classification for carbonaceous particles, reference samples of fly-ash needed from 32 power station ashes were obtained, from the Central Electricity Generating Board, the South of Scotland Electricity Board, the Northern Ireland Electricity Board and the Electricity Supply Board of Ireland. Of these ashes, 23 were from coal-fired stations, 7 from oil and 2 from peat.

Ten stations were selected, 5 coal and 5 oil (one peat station was also included in the characterisation study is to show that the scheme can be extended to include other fuel types) to produce a reference data set of particle chemistries 'typical' of these fuel types. A sample of the reference material from each of these stations was put through the carbonaceous particle extraction technique. This was for two reasons, firstly, so that the reference material had been subjected to the same chemical treatments as the sediment extracted particles, and secondly to remove the inorganic ash spheres from the coal ash reference samples.

EDS was performed on these samples as described by Watt (1990), for the following elements: Na, Mg, Al, Si, S, P, Cl, Ca, K, Ti, V, Cr, Mn, Fe, Ni, Cu and Zn. Outliers were removed where extreme values occurred for several elements on a single particle. These were often particles where the total X-ray count was very low showing them to be 'lacy' particles (Lightman & Street, 1968) *i.e.* where very little surface area remains after combustion. Once these outliers had been removed from the analysis, this left over 5500 particles upon which to base a classification scheme.

Characterisation

Principal components analysis (PCA) and stepwise discriminant analysis (SDA) were used to determine the most important elements in the separation. PCA transforms the original data into a series of principal components. The first principal component is the line of maximum variability passing through the data cluster, the second principal component is the line of maximum variability perpendicular to the first and so on. Each principal component has a loading from each element, and so by seeing which elements have the highest loading it is possible to see which elements explain most of the variance in the data. Most of the variance was found to be between the two fuel types and so these elements are also the ones important in the coal/oil separation.

The two most important elements were found to be sulphur and aluminium and these are highly negatively correlated (Pearson's coefficient = – 0.694). Sulphur has a slightly higher loading than aluminium, and when the 'fossil' particles (*i.e.* those extracted from sediments) were added to the analysis the difference between the loadings was increased. This seems to suggest that sulphur is the more important element when trying to separate fossil coal and oil particles.

SDA can also be used in fuel type separations by producing a ranking of the most important elements that separate the two groups. The ranking achieved for coal and oil particles is as follows:

S > V > Fe > Mg > Cl > Ti > Cr > Al > K > Si > Na > Mn > Ni > P > Zn > Ca > Cu,

and little improvement is made to the number of particles correctly allocated after the first 6 elements have been added to the classification function. From the results of these analyses, it was concluded that the most efficient fuel type separation was achieved when the first six elements from this ranking were included in a linear discriminant function (LDF) produced from discriminant function analysis (DFA). This correctly allocates over 94% of the particles.

Each particle is also given a probability of allocation to each fuel type. A threshold for this probability can be set so that only those particles with a probability higher than the threshold are allocated to fuel a type. Those falling below the threshold go into a group called 'unclassified'. This procedure increases the confidence that unknown particles allocated to a fuel type have been allocated correctly. If the threshold is set too low, many misallocated particles remain in the analysis, and if set too high, many correctly allocated particles fall into the unclassified group. The threshold setting which removes the highest number of misallocated particles from the analysis whilst minimising the number of legitimate particles falling into the unclassified group was found to be 0.8. (*i.e.* only those particles with a probability of allocation > 0.8 will be allocated a fuel type.) Using the LDF containing S, V, Fe, Mg, Cl & Ti only, and a probability threshold of 0.8, of the 94% of the particles allocated a fuel type, over 97% of them are allocated correctly.

Spatial distribution of characterised particles

To study the impact of the combustion of different fossil-fuels on Scottish lochs, the carbonaceous particles extracted from the surface levels of 17 Scottish sediment cores were characterised using the methods described above.

As might be expected in a country where there are few oil-fired power stations, the results show that coal is the dominant fossil-fuel, with over 87% of the characterised particles allocated to coal at every site. However, the east coast and south-west of Scotland appear to be areas where there is a higher oil influence. Figure 1 shows a contour map of percentage oil particles interpolated from the above data. The high point around Glasgow is due to the value for Loch Tinker. This may be higher than expected, because it is from a core taken in June 1985, at a time when the Inverkip oil-fired power station in Glasgow was still functional, only 40 km away. Inverkip closed down in 1987 and it may be that characterisation of particles extracted from the surface level of a core taken from Loch Tinker now, would show a lower oil value. Apart from these areas there appears to be very little oil influence (less than 3%) over the rest of the country.

The oil particles deposited in the south-west probably have their origins outside of the country. Oil percentage contours decline to the south of Galloway with the value for Whitfield Lough in the North Pennines being only 0.22% and this

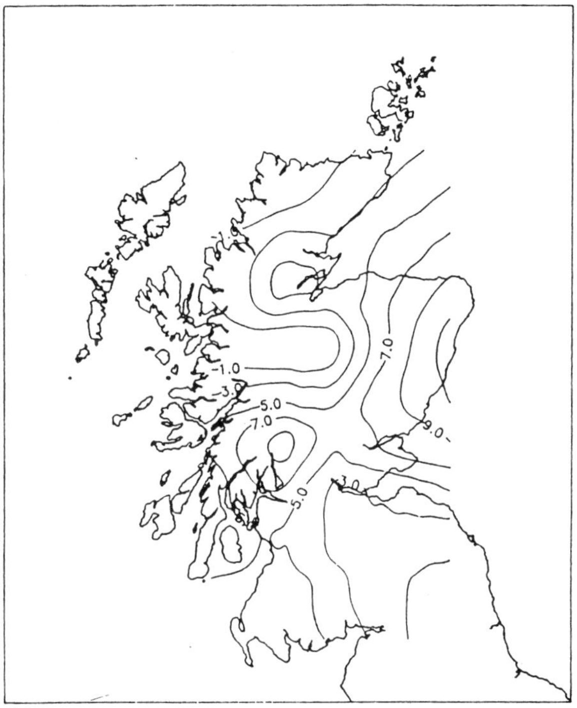

Fig. 1. Percentage of carbonaceous particles of oil origin in surface sediments from Scottish lochs.

suggests that they are not being transported northwards from England. The furthest north oil-fired power station in England is Ince 'B' on Merseyside about 200 km away. This is not a major plant and so the sources for the south-western oil particles are most likely to be the oil-fired power stations at Coolkeragh and Ballylumford in Northern Ireland.

Sources for the east coast oil particles are likely to be from the Peterhead oil-fired power station near Aberdeen. It should be remembered however, that these results are only generated by 17 points, and although there are several sites with 'high' oil in each of these regions, the map is far from conclusive. A further 20 sites concentrated around these two areas are in the process of being

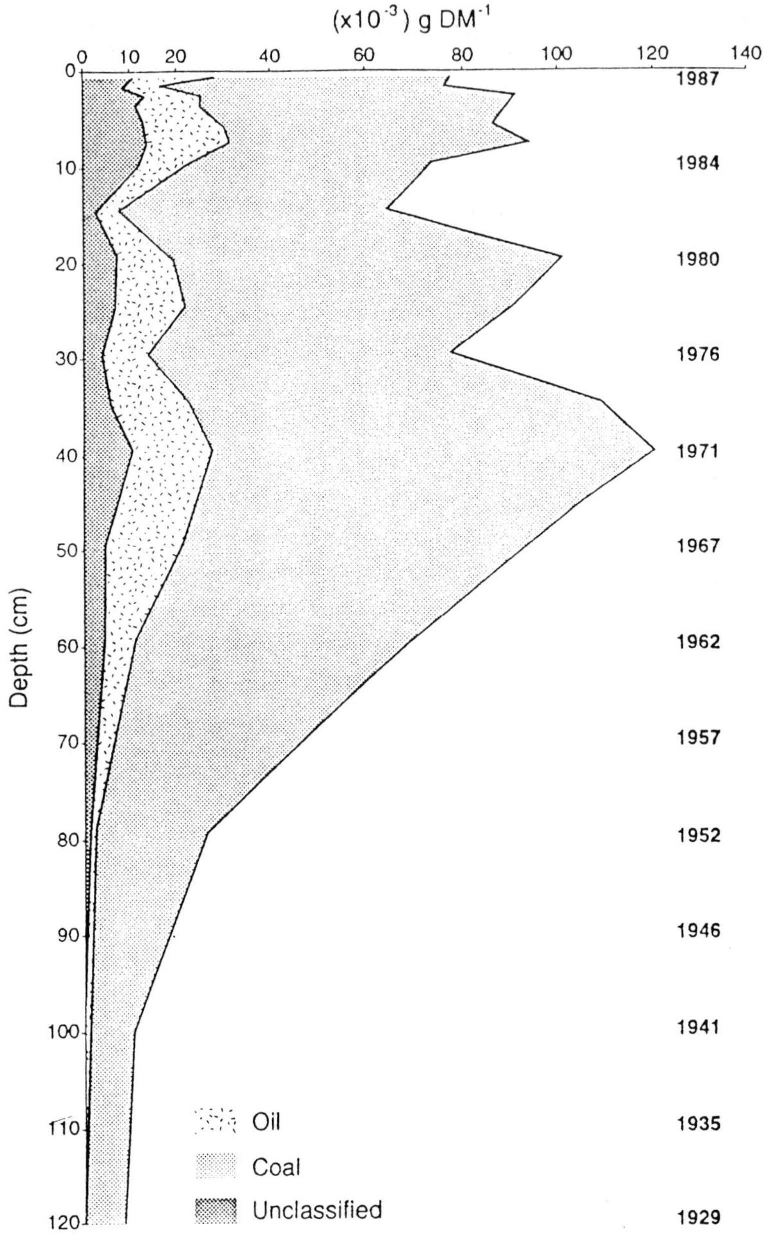

Fig. 2. The carbonaceous particle concentration profile for the Men's Bathing Pond, Hampstead Heath, divided into coal, oil and unclassified groups.

characterised and it is hoped that these extra sites will help confirm these results.

Temporal distribution of characterised particles

The characterisation technique was then applied to carbonaceous particles extracted from a sediment core taken from the Men's Bathing Pond on Hampstead Heath in North London. This site was selected as there were many particles present at all levels, which maximises EDS efficiency and also there has been a change from exclusively coal to some oil burning in this region since the 1920's, (*i.e.* the base of the core). 7504 carbonaceous particles from 18 levels were characterised as described above. The results are shown in Table 1 and Fig. 2.

These results fit very well with the known combustion histories of coal and oil. Before the 1920's fossil-fuel consumption in Britain was almost exclusively coal based, although from this time onwards the use of oil increased rapidly from only 3 million tonnes in 1920 to 160 million tonnes in 1972. This was principally due to an influx of cheap fuel oil after the Second World War and as a response to this the first major power station specifically designed to be oil-fired was opened in 1952 on the River Thames at Bankside. Oil consumption increased rapidly in the late 1950's/early 1960's and continued to increase until the oil crisis in 1974. Since then, there has been a general decrease in oil consumption, with the exception of the miners strike in 1984/85, and coal has remained the dominant fuel throughout. Despite the limited number of data points, most of these events can be seen in Fig. 2. There are very few oil particles until the 1950's, a sharp increase at the end of that decade, and a steady increase until the mid-1970's, when the number generally decreases except for a peak at about 1984. This shows that it may be possible to add extra dates to a carbonaceous particle profile using the introduction of fuels as new dating horizons. This would supplement dates obtained from carbonaceous particle concentration profiles (Renberg & Wik, 1984; Rose, 1991)

Table 1. Fuel type allocation for the Hampstead Health core.

Depth (cm)	Date	% coal	% oil	% unclassified
0–1	1987	63.8	22.8	13.4
1–2	1987	79.0	9.3	11.7
2–3	1986	72.6	13.0	14.4
3–4	1986	72.5	14.7	12.8
5–6	1986	65.4	19.3	15.3
7–8	1985	67.1	18.4	14.5
9–10	1984	68.9	14.6	16.5
14–15	1982	88.5	7.4	4.1
19–20	1980	80.9	11.9	7.2
24–25	1978	76.1	17.1	6.8
29–30	1976	82.6	12.4	5.0
34–35	1974	79.6	15.6	4.8
39–40	1971	77.6	14.1	8.3
49–50	1967	78.0	17.4	4.6
59–60	1962	84.7	9.3	6.0
79–80	1952	92.4	3.3	4.3
99–100	1941	94.7	2.1	3.2
119–120	1929	99.8	0.0	0.2

Conclusions

Carbonaceous particles from coal and oil combustion can be readily separated using the elemental chemistry data produced by EDS analysis of the particle surfaces. A LDF involving six elements enables this separation to be done with over 97% of the particles allocated to a fuel type being allocated correctly. This technique has been applied to both surface sediments to study spatial distribution and to a dated sediment core to study changes through time and the results correlate well with known combustion distributions and histories and so can be used to add additional dates to sediment cores.

Other work (Rose, 1991) has shown that the possibility of taking this particle characterisation a stage further to power station level seems unlikely, but extending it to include other fuel types commonly used in other parts of Europe seems highly plausible. Only a preliminary study using peat data has been undertaken, but the results suggest that an effective characterisation including peat could be produced. The work of Mejstrik & Svacha (1988) shows that fly-ash produced from power stations burning brown coal and lignite in Czechoslovakia is enriched in Co, Zn, Cr, Ni and Cd. If this is true of all brown coal and lignite ashes then there should be few problems in including these fuel types in a future characterisation scheme. It may also be possible to include particles derived from oil shale burning (e.g. in Estonia and N.W. Russia) as these ashes are likely to be very high in calcium (Veiderma, pers. comm.). Oil shale combustion affects environments in Sweden, Finland as well as Russia and the northern Baltic states.

The major pollutants of mountain lakes are atmospheric in origin and so the inclusion of these other fossil-fuel types into the classification scheme will make the characterisation of carbonaceous particles a powerful technique for identifying possible sources of these pollutants in these and other lakes over the whole of Europe.

Acknowledgements

I would like to thank John Watt of Particle Characterisation Services at Imperial College London for the EDS analyses, and Peter Appleby at the University of Liverpool for the core dating.

I am very grateful to Steve Juggins for all his help with the multivariate statistics and computing.

Thanks are also due to the members of the Palaeoecology Research Unit at University College London for help with the field work, and to Rick Battarbee for his supervision and useful comments on the manuscript and throughout the project.

This work was partially funded by the Central Electricity Generating Board and the Department of the Environment.

References

Appleby, P. G., P. Nolan, D. W. Gifford, M. J. Godfrey, F. Oldfield, N. J. Anderson & R. W. Battarbee, 1986. ^{210}Pb dating by low background gamma counting. Hydrobiologia 141/Dev. Hydrobiol. 36: 21–27.

Battarbee, R. W., N. J. Anderson, P. J. Appleby, R. J. Flower, S. C. Fritz, E. Y. Haworth, S. Higgitt, V. J. Jones, A. Kreiser, M. A. R. Munro, J. Natkanski, F. Oldfield, S. T. Patrick, N. G. Richardson, B. Rippey, A. C. Stevenson, 1988. Lake Acidification in the United Kingdom 1800–1986. Ensis Publishing. London: 68 pp.

Goldberg, E. D., 1985. Black carbon in the environment: Properties and distribution. Wiley Interscience Publication. New York: 198 pp.

Griffin, J. J. & E. D. Goldberg, 1981. Sphericity as a characteristic of solids from fossil fuel burning in a Lake Michigan sediment. Geochim. et Cosmochim. Acta. 45: 763–769.

Kajak, Z., 1966. Field experiment in studies on benthos density of some Mazurian lakes. Gewäss und Abwäss. 41/42: 150–158.

Lightman, P. & P. J. Street, 1968. Microscopical examination of heat treated pulverised coal particles. Fuel 47: 7–28.

Livingstone, D. A., 1955. A lightweight piston sampler for lake deposits. Ecology 36: 137–139.

Mackereth, F. J. H., 1969. A short core sampler for subaqueous deposits. Limnol. Oceanogr. 14: 145–151.

Mejstrik, V. & J. Svacha, 1988. The fallout of particles in the vicinity of coal-fired power plants in Czechoslovakia. Sci. Tot. Env. 72: 43–55.

Raask, E., 1984. Creation, capture and coalescence of mineral species in coal flames. J. Inst. Energy. 57: 231–239.

Renberg, I & M. Wik, 1984. Dating of recent lake sediments by soot particle counting. Verh. int. Ver. Limnol. 22: 712–718.

Rose, N. L., 1990. A method for the extraction of carbonaceous particles from lake sediment. J. Paleolimnol. 3: 45–53.

Rose, N. L., 1991. Fly-ash particles in lake sediments: Extraction, characterisation and distribution. Unpublished PhD thesis. University of London.

Watt, J., 1990. Automated feature analysis in the scanning electron microscope. Microscopy & Analysis. Issue 15.

Hydrobiologia **274**: 133–142, 1994.
J. Fott (ed.), Limnology of Mountain Lakes.

Autotrophic picoplankton community dynamics in a pre-alpine lake in British Columbia, Canada

J. G. Stockner & K. S. Shortreed
Department of Fisheries and Oceans, West Vancouver Laboratory, 4160 Marine Drive, West Vancouver, British Columbia V7V 1N6

Key words: Autotrophic picoplankton, mountain lake, phytoplankton, nutrients, grazing.

Abstract

Autotrophic picoplankton (APP) were studied in Chilko Lake, a large, deep ultra-oligotrophic pre-alpine lake (elevation: 1172 m) in the south central coast mountains of British Columbia. Data from 1985 (untreated) and 1990 (treated) were used to compare and contrast APP community response to a whole-lake fertilization experiment. The APP communities of Chilko Lake were dominated by the coccoid cyanobacteria *Synechococcus* and its colonial morph which comprised about 99% of the APP community of Chilko Lake. *Chlorella*-like eukaryotic picoplankters and small cyanobacteria were rare, comprising <1% of the APP community. In 1990 autotrophic picoplankters contributed an average of 73% to total chlorophyll, and 54% to total photosynthesis. Average APP abundance ranged from lows of 4,000–5,000 cells ml^{-1} in winter and spring to highs of 50000–150000 cells ml^{-1} in early August with no apparent autumnal increase. APP populations were uniformly distributed in the epilimnion, but during calm periods in August often formed a peak near the metalimnion/hypolimnion boundary. Seasonal and vertical distribution patterns of APP showed little relation to temperature or to light. When nutrients were added to the lake in 1990, APP populations doubled within 3 wk of addition and average abundance (6.16×10^4cells $\cdot$ ml^{-1}) was twice 1985 APP numbers. Bottom-up control by scarce nutrient supplies is considered the primary factor regulating community composition and abundance during the initial population growth phase (June, July) with top-down control by grazing during nutrient co-limitation periods when the epilimnion is deplete of both nitrogen and phosphorus (August, September).

Introduction

It has only been within the past decade that autotrophic picoplankton (APP) have been recognized as important components of lake phytoplankton communities (Stockner & Antia, 1986; Stockner & Porter, 1988). Recent studies have shown that they can contribute significantly to photosynthetic carbon production and to biomass in a wide spectrum of lakes (Stockner, 1991). With their small size, positive buoyancy, photopigment complement and potential for rapid growth these minute, single-celled phototrophs appear especially well adapted to the nutrient deplete euphotic zones of oligotrophic lakes where they are often the most abundant algal group. However, they can also make large contributions to phytoplankton community metabolism in eutrophic and even hypereutrophic ecosystems (Ulhmann, 1966; Sondergaard, 1991; Voros *et al.*, 1991). The seasonal and vertical distribution and abundance of APP has been described from lakes throughout the world (Stockner & Shortreed, 1991; Burns & Stockner, 1991; Sondergaard, 1991; Nagata, 1986). Their physiological requirements (Suttle & Harrison, 1988; Suttle *et al.*, 1988; Wehr, 1991) and their contribution to food webs in lacustrine systems (Stockner, 1987; Weisse, 1988; Nagata, 1988; Stockner & Shortreed, 1989) are also now better understood. However, there is a paucity of reports of APP abundance and distribution in mountain lakes in North America. The primary objectives of this study were: (1) to report on the composition, seasonal abundance and distribution of the APP community in an ultra-oligotrophic pre-alpine lake in British Columbia, and (2) to describe the response of this community to a nutrient addition bioassay and a lake fertilization experiment.

Description of study lake

Chilko Lake (51° 20′ N, 124° 05′ W) is situated at an elevation of 1172 m at the boundary of the coast mountains and the interior plateau in southern British Columbia (Fig. 1). The climate is continental, with warm, dry summers and long, cold winters. The lake is located in the Cariboo aspen-lodgepole pine and subalpine Engelmann spruce-subalpine fir biogeoclimatic zones (Farley, 1979). Annual precipitation averages < 100 cm. With an area of 200 km^2, it is one of the larger lakes of the Fraser River drainage basin. The steep-sided nature of the surrounding terrain and of the lake itself results in an extremely limited littoral area. Only at a relatively small (app. 1.5 km^2) area near the outlet of the lake is the littoral zone more than a narrow band. Despite the cold temperatures, complete ice cover does not occur on the lake in most winters. We have elsewhere (Stockner & Shortreed, 1991) suggested that Chilko Lake be classified as cool monomictic (one period of circulation each winter at <4 °C), since in most years it is neither dimictic nor warm monomictic. Mean depth of the lake is 123 m, maximum depths exceed 300 m, and during our study (1984–1990) theoretical water residence times ranged from 17 to 25 yr. During summer, the southern portion of Chilko Lake receives several glacially turbid inflows, with the result that water clarity tends to decrease both spatially from north to south and seasonally from May to August. The orientation of Chilko Lake at the edge of the coast mountains results in frequent strong southerly catabatic winds. These southerly winds exert a very strong influence on the circulation patterns and thermal structure of the lake. Chilko Lake is an important nursery area for anadromous and economically valuable sockeye salmon (*Oncorhynchus nerka*). Adult sockeye escapements to the lake exceed 1 million in some years, resulting in a large recruitment of planktivorous juvenile sockeye salmon to the littoral and limnetic areas of the lake. These juveniles spend 1 yr (a small minority spend 2 yr) in the lake prior to seaward migration.

Methods

Lake fertilization experiments were carried out in 1990 with an aqueous solution of ammonium nitrate and ammonium phosphate applied to the north central basin at weekly intervals for a period of 14 wk. The application rate was 4 mg $P \cdot m^{-2}\ wk^{-1}$ at a molar N:P ratio of 25:1. The technique has been widely utilized in British Columbia for enhancement of sockeye salmon (*Oncorhynchus nerka*) and has been described in detail elsewhere (Stockner, 1987).

Since 1984 we have sampled a number of locations in the lake with frequencies ranging from weekly to monthly. For the purposes of this presentation we focus on results from station 11, which is located 16 km from the northern, or outlet end of the lake, and was sampled once weekly in 1985 and 1990 from May until October (Fig. 1). The station was also located within the fertilized basin of the lake during the 1990 fertilization experiment. We used Applied Microsystems Ltd. CTD meters (Models CTD-12 and STD-12) to measure conductivity, temperature and depth. Photosynthetic photon flux density (PPFD: 400–

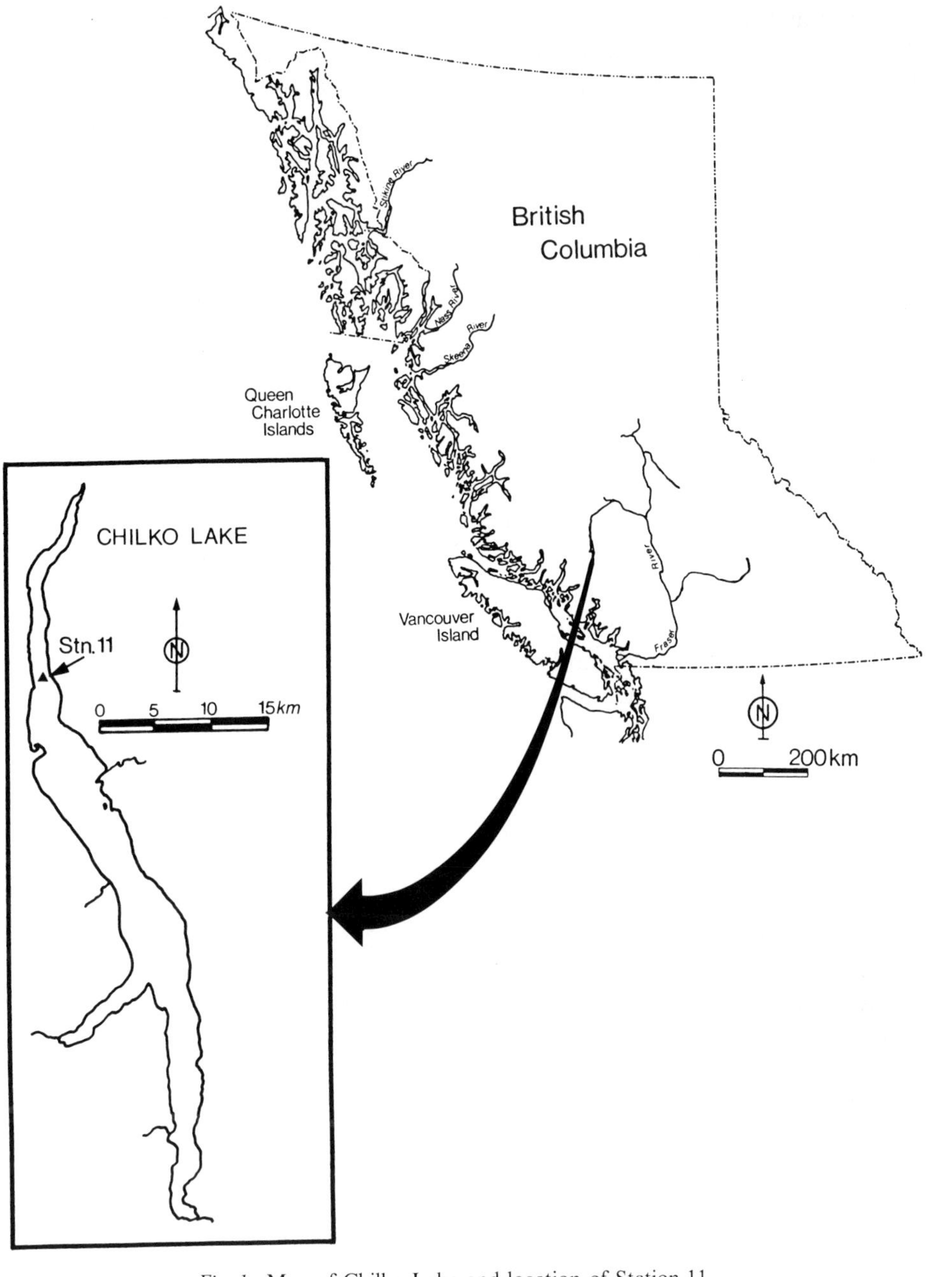

Fig. 1. Map of Chilko Lake and location of Station 11.

700 nm) was determined using a Li-Cor Model 185A light meter equipped with a Model 192S underwater quantum sensor, and compensation depths and vertical light extinction coefficients were calculated. Daily solar radiation was determined using either a Li-Cor Model 550 printing integrator (1985) or a Li-Cor Model LI-1000 data logger (1990), each equipped with a Model 190S quantum sensor.

An opaque 6-L Van Dorn water sampler was used to collect all water samples, usually between 0900 and 1100 h. Water from pre-selected depths from the surface to the hypolimnion or compensation depth was collected for analysis of chemical variables and for phytoplankton biomass, species composition, and *in situ* production rates.

All reported chemical analyses were done following methods described by Stephens and Brandstaetter (1983) and Stockner & Shortreed (1985). Water for dissolved nutrient analyses were filtered through ashed Whatman GF/F filters which were rinsed with both distilled deionized water (DDW) and sample immediately prior to filtration. Filtered water was stored in clean glass and polyethylene bottles and kept cold and dark prior to analysis. Nitrate was determined by converting to nitrite by cadmium reduction. Water for total phosphorus (TP) analysis was collected in clean, screw-capped, glass test tubes, and TP was later determined using a molybdenum blue method after persulfate digestion.

Samples for total chlorophyll *a* determination were collected on Millipore AA filters (0.8 μm nominal pore size), desiccated, frozen, and later analyzed fluorometrically after maceration in 90% acetone. APP chlorophyll was determined indirectly in 1990 by analyzing chlorophyll on 2.0 μm Nuclepore filters, and subtracting from the total chlorophyll analysis. *In situ* photosynthetic rates were determined by filling 125-ml glass bottles with water from each sampling depth, inoculating with a sodium bicarbonate solution containing ^{14}C, and incubating for 1.5–2 h at the original sampling depths. Incubations usually commenced between 0900 and 1000 h. After the incubations, samples were filtered onto 0.2 μm and 2.0 μm pore size Nuclepore filters, placed in scintillation vials containing a tissue solubilizer and scintillation cocktail, and counted in a Packard scintillation counter. Dissolved inorganic carbon (DIC) concentrations were determined using the potentiometric method of APHA (1976). Production rates were calculated using Strickland's (1960) equation.

Phytoplankton larger than the picoplankton size fraction (> 2 μm maximum dimension) were identified and enumerated at 560 × and 1200 × using a Wild inverted microscope equipped with phase optics. APP were enumerated using methods describe by MacIsaac and Stockner(1985). They were collected on a 0.2 μm pore Nuclepore filter stained with Irgalan Black at a vacuum pressure < 450 Pa. Enumerations were done using a Zeiss epifluorescence microscope equipped with a 397 nm longwave-pass exciter filter and a 560 nm shortwave-pass exciter filter, a 580 nm beam-splitter mirror and a 590 nm longwave pass barrier filter. APP were classified into four categories using autofluorescence color, presence or absence of a chloroplast, and aggregation (aggregates of > 4 cells were labelled colonial; eukaryotic picoplankton were not seen in colonies). USYN are unicellular, orange (phycoerythrin) fluorescing cyanobacteria (*Synechococcus*). CSYN are the same except they occur in colonial form. RCYN are small (< 1 μm) red (phycocyanin) fluorescing cells without a visible chloroplast. REUK are larger (1–2 μm) red fluorescing cells with a visible chloroplast (eukaryotes). The carbon content of the phytoplankton community was estimated using the equation of Fahnenstiel *et al.* (1991) for APP and Strathmann's (1967) equation for larger phytoplankton.

A nutrient bioassay was carried out from August 9–20, 1985. Replicate 4-l Erlenmeyer flasks were used for each of the 6 treatments, which were: 1. control (C), 2. 170 $\mu g\ N \cdot l^{-1}$ (N), 3. 15 $\mu g\ P \cdot l^{-1}$ (P), 4. 5 $\mu g\ P \cdot l^{-1}$ and 57 $\mu g\ N \cdot l^{-1}$ (N5P), 5. 10 $\mu g\ P \cdot l^{-1}$ and 113 $\mu g\ N \cdot l^{-1}$ (N10P), and 6. 15 $\mu g\ P \cdot l^{-1}$ and 170 $\mu g\ N \cdot l^{-1}$ (N15P). Nutrients consisted of NH_4NO_3 and K_2HPO_4 and were added once only to the flasks on the second day of the 12-d bioassay. Ambient nutrient concentrations prior to spiking the flasks were 1 $\mu g\ N \cdot l^{-1}$ nitrate and 2 $\mu g \cdot l^{-1}$ total phosphorus. In flasks which received both nitrogen (N) and phosphorus (P), the N:P molar ratio of added nutrients was 25:1. The bioassay was conducted outdoors. Temperature was maintained at ambient levels by submerging the flasks in the Chilko River, which drains the lake. Light levels were maintained at approximately 10% of surface intensity by positioning several layers of mesh screens above the flasks. Water for the bioassay was collected on August 8 from a depth of 10 m at station 11. Flasks were filled immediately and nutrients were added on August 10. Flasks were

Table 1. Differences in salient physical and chemical variables in 1985 (unfertilized) and in 1990 (fertilized).

Year	Euphot. depth (m)	Secchi depth (m)	Mean epi. temp. (°C)	Total P $\mu g \cdot l^{-1}$	Part. P $\mu g \cdot l^{-1}$	NO_3-N $\mu g\ N \cdot l^{-1}$	Tot. diss. solids $mg \cdot l^{-1}$	Silicate $mg\ Si \cdot l^{-1}$	pH
1985	16.6	4.3	10.6	2.4	1.4	12	33	0.95	6.9
1990	19.9	6.6	12.3	3.7	1.8	11	34	1.19	7.5

gently swirled every day and were sampled every second day.

Results

Physical-chemical variables

Seasonal average mean epilimnetic temperatures were 10.6 °C in 1985 and 12.3 °C in 1990 (Table 1). Epilimnion depths were variable (10–20 m), but thermal stratification generally developed by late June and continued until mid-October. In 1985 euphotic depths averaged 17 m and ranged from 15 to 23 m. Euphotic zones averaged 20 m in 1990 and ranged from 11 to 45 m (all averages are calculated from early June to early October). Average epilimnetic pH was 6.9 in 1985 and 7.5 in 1990; total dissolved solids (TDS) averaged 33 $mg \cdot l^{-1}$. TP concentrations averaged 2.4 $\mu g \cdot l^{-1}$ in 1985 and 3.7 in 1990. Nitrate levels averaged 12 $\mu g\ N \cdot l^{-1}$ in both study years and ranged from 20–25 $\mu g \cdot l^{-1}$ at overturn.

Biological variables

Average chlorophyll concentrations were 0.86 $\mu g \cdot l^{-1}$ in 1985 and 0.69 $\mu g \cdot l^{-1}$ in 1990 (Table 2). Mean epilimnetic chlorophyll concentrations were generally lowest in spring and highest in late summer or fall; they ranged from 0.24–1.54 $\mu g \cdot l^{-1}$. In 1990 the seasonal average epilimnetic contribution of APP to total chlorophyll was 73% and to total photosynthesis 54%. Picoplankton comprised an average of 2% of phytoplankton biomass (carbon) in 1985 and 29% in 1990 (Table 3).

The predominant picoplankter was the coccoid cyanobacteria *Synechococcus*, ranging in size from 0.8–1.6 μm with phycoerythrin as the predominant phycobilin pigment (yellow-orange fluorescence). USYN comprised 89% of the total population in both study years, with CSYN comprising the remainder (Table 3). RCYN and REUK were present in extremely low numbers and always comprised <1% of the total population. Seasonal average APP numbers were $3.07 \times 10^4 \cdot ml^{-1}$ in 1985 and $6.61 \times 10^4 \cdot ml^{-1}$ in 1990; seasonal maxima occurred in early August (Table 3, Fig. 2).

APP vertical distribution was most often uniform within the epilimnion but when periods of protracted calm weather occurred metalimnetic peaks often formed (Fig. 3). In both study years seasonal maxima of hypolimnetic APP populations occurred in early October. Seasonal aver-

Table 2. Differences in selected biological variables between 1985 and 1990. Data are seasonal averages of mean epilimnetic data.

Year	Bacteria numbers $\times 10^6 \cdot ml^{-1}$	Total chlorophyll $\mu g \cdot l^{-1}$	% Picopl. chl.	Mean daily photosyn. rate $mg\ C \cdot m^{-2} \cdot day^{-1}$	% Picopl. photosyn.
1985	1.09	0.86	no data	111	no data
1990	0.86	0.69	73	131	54

Table 3. Changes in seasonal (June–September) average epilimnetic picoplankton numbers in a fertilized and unfertilized year. Numbers in brackets are phytoplankton carbon in pg · ml^{-1}.

Year	Phytoplankton numbers ($\times 10^3 \cdot ml^{-1}$)			
	USYN	CSYN	Nanopl.	Micropl.
1985	27.5 (1.7)	3.2 (0.6)	7.2 (34.2)	1.6 (63.9)
1990	58.9 (13.3)	7.2 (1.4)	7.1 (16.3)	0.3 (20.1)

ages of hypolimnetic APP were $8.25 \times 10^3 \cdot ml^{-1}$ in 1985 and $2.64 \times 10^4 \cdot ml^{-1}$ in 1990. USYN was the predominant species at all depths with CSYN more abundant in the near-surface waters.

After commencement of fertilization APP rapidly increased to an early August peak of over 150000 · ml^{-1}, more than double the 1985 seasonal maxima (Fig. 2). Neither REUK nor RCYN increased with treatment.

During the nutrient bioassay APP numbers slowly declined in the C, N, and P treatment from the starting concentration of $8.8 \times 10^4 \cdot ml^{-1}$ until lows of $4.8 \times 10^4 \cdot ml^{-1}$, $1.8 \times 10^4 \cdot ml^{-1}$, and $2.0 \times 10^4 \cdot ml^{-1}$, respectively were reached by the end of the bioassay (Table 4). CSYN comprised < 10% in each of the three treatments. Greatest APP response occurred in the N5P treatment,

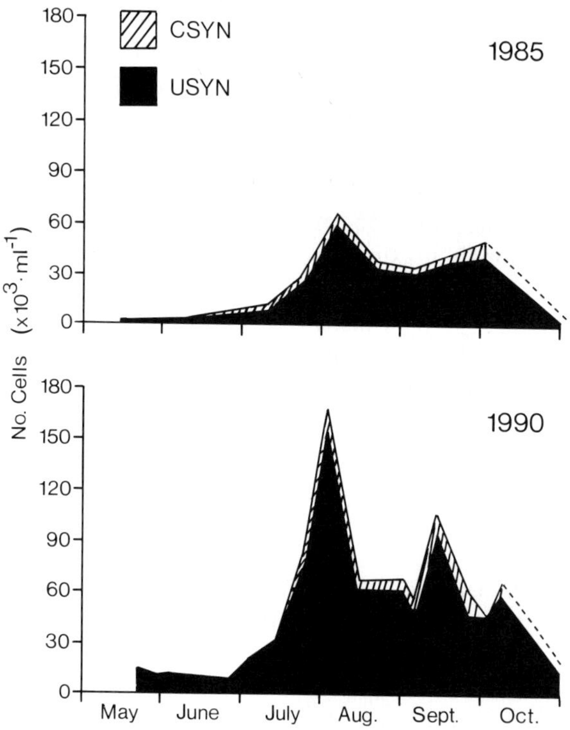

Fig. 2. Seasonal abundance of unicellular *Synechococcus* (USYN) and the colonial morph of *Synechococcus* (CSYN) at Station 11, Chilko Lake in 1985 and 1990.

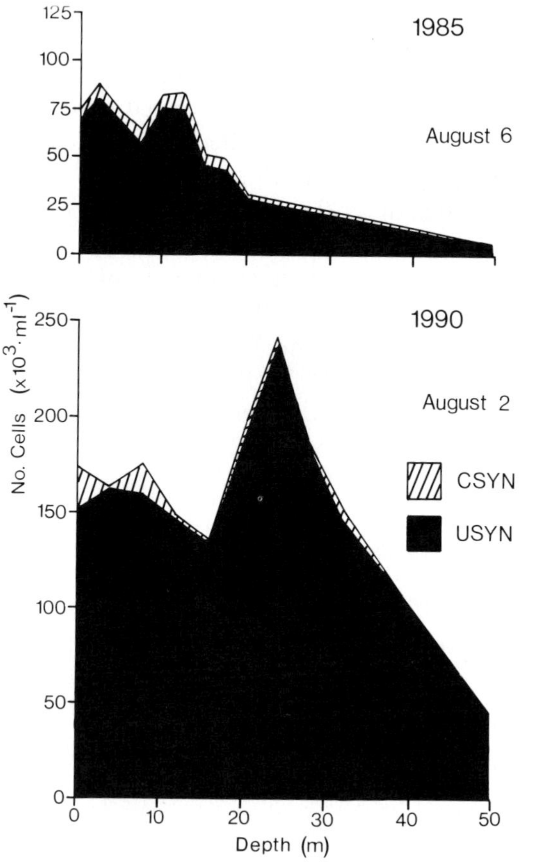

Fig. 3. Vertical distribution of unicellular *Synechococcus* (USYN) and the colonial morph of *Synechococcus* (CSYN) on August 6, 1985, and August 2, 1990 at Station 11, Chilko Lake.

Table 4. Final yields from a nutrient bioassay carried out in August 1985. Data were collected 10 days after spiking with nutrients. Prior to nutrient additions, phytoplankton numbers (in thousands) were 87.9 for USYN, 0.5 for CSYN, 2.5 for nanoplankton, and 0.4 for microplankton.

Treatment	Phytoplankton numbers ($\times 10^3 \cdot ml^{-1}$)			
	USYN	CSYN	Nanoplankton	Microplankton
Control	46.7	1.0	4.4	1.3
170 $\mu g\ N \cdot l^{-1}$	16.9	0.8	4.3	1.3
15 $\mu g\ P \cdot l^{-1}$	18.4	1.9	4.1	1.6
5 $\mu g\ P \cdot l^{-1}$ + 57 $\mu g\ N \cdot l^{-1}$	154.6	229.0	6.7	1.8
10 $\mu g\ P \cdot l^{-1}$ + 113 $\mu g\ N \cdot l^{-1}$	138.5	79.4	8.9	4.1
15 $\mu g\ P \cdot l^{-1}$ + 170 $\mu g\ N \cdot l^{-1}$	90.6	79.2	6.7	5.3

where APP numbers reached $3.8 \times 10^5 \cdot ml^{-1}$, with 60% as CSYN. Increasing nutrient loads produced a lesser APP response, as numbers were $2.2 \times 10^5 \cdot ml^{-1}$ in the N10P treatment and $1.7 \times 10^5 \cdot ml^{-1}$ in the N15P treatment. Greatest nanoplankton response occurred in the N10P treatment and the greatest microplankton response in the N15P treatment.

Discussion

Community composition

Unicellular *Synechococcus* spp., (USYN) are the most abundant phototrophs in temperate lakes and oceans worldwide (Stockner, 1991; Glover, 1985). Not surprisingly, they were dominant in the Chilko Lake APP community as well. The colonial morph of *Synechococcus* (CSYN) was also present in both study years. Colonial APP are infrequently reported in salt water, but in lakes they can become a common component of the APP community, with greatest densities reported in more productive oligotrophic and mesotrophic lakes (Stockner & Shortreed, 1991; Fahnenstiel *et al.*, 1986). These APP colonies or loosely aggregated cell associations are thought to confer some advantage over single cells at times of severe nutrient scarcity by providing 'microzones' or sites of rapid nutrient and DOC recycling (Goldman, 1984; Sieburth 1988; Klut & Stockner, 1991). Others have suggested that the colonial morph may confer some refuge from grazing by microflagellates and ciliates and other microzooplankton (Güde, 1989; Pedros-Alio, 1989). Small eukaryotic APP (REUK) were rare in Chilko Lake, perhaps because of higher pH and alkalinity levels in Chilko Lake than in coastal lakes where eukaryotes are more abundant (Stockner & Shortreed, 1991). Small (0.4–1.0 μm), coccoid (phycocyanin containing) cyanobacteria (RCYN), are present and occasionally abundant in some B.C. coastal dystrophic lakes (Stockner & Shortreed, 1991), but were quite rare in Chilko Lake.

Seasonal abundance

Average epilimnetic densities (30–60 000 cells $\cdot ml^{-1}$) of APP in Chilko Lake were typical of temperate ultraoligotrophic lakes worldwide (Stockner, 1991). However patterns of APP seasonal abundance differed from the more common pattern seen in other temperate lakes in British Columbia (Stockner & Shortreed, 1991) and elsewhere (Pick & Caron, 1987; Weisse & Kenter, 1991). Population increases in Chilko commenced in mid-July, reaching peak abundance in early August, and exhibited no prominent autumnal increases. In contrast, in most lower elevation coastal and interior B.C. lakes APP begin to increase in June and attain maximum abundance in September and October (Stockner, 1987; Stockner & Shortreed, 1991). Because of the large size,

dystric planosols prevail in the upper plain area of the watershed whereas podzols dominate in the steeper catchment slopes (Bächle, 1991). The main vegetation of the 67 ha watershed consists of Norway spruce (*Picea abies* (L.) Karst.) and fir (*Abies alba* Mill.) which both show severe damages induced by acidic deposition. The cirque lake was formed during the Würm glaciation period and has a maximum depth of 8 m, a volume of 65 000 m^3, an open water surface area of 2.0 ha, and an estimated theoretical mean water retention time of one month (Thies, 1990). The original shore line of the lake is formed by a floating *Sphagnum* peat mat. Due to the construction of a small dam about one century ago a shallow (2 m deep) peripheral part (without floating *Sphagnum* peat mat at its margin to the forest) was formed around the former lake, where the water lily *Nuphar lutea* Sm. could spread out due to the shallowness of the water (Thies, 1987). Lake Huzenbach is a dystrophic headwater lake which is potentially meromictic (Thies, 1991a). Its actual trophic state is mesotrophic (Schröder, 1991).

Materials and methods

Field work & laboratory analysis

During the 1988/1990 investigation period, lake inflows, outflow, and lake vertical profiles were sampled weekly. In the littoral zone and in lake vertical profiles, dissolved oxygen was measured in situ with a WTW Oxi96 oximeter (Wiss. Techn. Werkstätten, Weilheim) equipped with 10 m cable and a battery operated stirrer. pH values were determined both in situ and in the laboratory with a WTW pH 91 meter with automatic temperature adjustment in the quiescent sample after calibrating the Ingold LoT-401 glass electrode (Ingold, Steinbach/Ts.) with Merck pH 7.00 and 4.00 buffer standards (Merck, Darmstadt). Conductivity and temperature were measured in situ with a WTW LF91 meter. Alkalinity was determined in the laboratory at the day of sampling according to Gran (1952) while titrating with 0.02 N HCl values of approx. 4.3, 4.0 and 3.6 consecutively. Lake inflows and outflow samples were analyzed biweekly and vertical profile samples (0, 2, 4, 6, 8 m) monthly for the following parameters after filtration with acid-washed cellulose acetate filters of 0.45 μm porosity (ME 25, Schleicher & Schuell). Samples for cation analyses were acidified (adding 0.5 vol% 4 N HNO_3 suprapur, Merck) after filtration. Na, K, Ca, and Mg were analyzed with a flame atomic absorption spectrometer Perkin-Elmer Model 430 (Perkin-Elmer, Bodenseewerk Überlingen). Al, Fe, and Mn were analyzed with a graphite furnace atomic absorption spectrometer Perkin-Elmer/Zeeman Model 3030. Speciation of Al, Fe and Mn was performed according to Driscoll (1984). Anions were analyzed with a Dionex 2020i ion chromatography (Dionex GmbH, Idstein) for Cl, NO_3 and SO_4. Dissolved organic carbon (DOC) samples were frozen after filtration and analyzed with a modified Maikak/Defor Unor 4 N TOC infrared gasanalyzer (Maihak, Hamburg; modified by Gröger & Obst and Wagner). SiO_2, NH_4, NO_2, PO_4, dissolved organic nitrogen (DON), total dissolved phosphorus (TDP), and UV absorption at 254 nm were analyzed with Perkin-Elmer Lambda2 UV/VIS and Zeiss-Elko II spectrometers using German standard methods (DEV, 1985) and related colorimetric methods (Institute of Lake Research, Konstanz & Langenargen, unpubl.). Water samples for cations, anions and nutrients were stored after sampling in acid-washed polyethylene vessels at about 4 °C in the dark for less than 48 hours prior to analysis. Bulk precipitation was sampled in the open and in the canopy throughfall with polyethylene gauges which were equipped with a device to minimize evaporation. The amount of precipitation was determined in a biweekly to monthly turn and samples were analyzed for conductivity, pH, alkalinity, Na, K, Ca, Mg, Al, Fe, Mn, Cl, NO_3, SO_4, DOC, UV 254 nm, NH_4 and TDP following the analytical procedures given above. Dataloggers (Microbit Informations-Systeme GmbH, Oberhausen) were installed in early 1989 in one spring inlet gauging station and in the outlet gauging station in order to record conductivity, pH, temperature, and water level every twenty minutes.

Plausibility tests & statistical analyses

Water analyses were checked by plausibility tests. Theoretically calculated conductivity was compared with measured conductivity and ion balances were calculated using PC software of Limnology Bureau Hoehn (Freiburg). Errors in raw data arising from data transcription, computer processing, etc. were detected and corrected by applying these methods. Statistical analyses were performed with SPSS-X software (SPSS Inc., Chicago, Illinois) applying Spearman rank correlations.

Results

Precipitation

Bulk deposition values for open precipitation and canopy throughfall in the hydrological years 1989 and 1990 are given in Table 1. The amount of open precipitation during the hydrological year 1989 (01.11.1988–31.10.1989) at Lake Huzenbach was 1352.9 mm. In the hydrological year 1990 (01.11.1989–31.10.1990) it was 1764.5 mm. Precipitation values of 1551 mm and 1643 mm were recorded at the weather-station Forbach-Herrenwies during the hydrological years 1989 and 1991, respectively (German Weather Service, 1992, 1993). This weather-station is situated at 764 m a.s.l. within 10 km linear distance northwest from Lake Huzenbach. The hydrological year 1989 was the driest year during the period 1980–1990, and pluriannual mean values of 1955 mm were reported for the 1931–1960 period and 1866 mm for the 1951–1980 period at this weather-station (W. Mikuteit, German Weather Service, Weather Bureau Freiburg, pers. com.). The canopy throughfall accountes for the major contribution of acidic deposition in the watershed of Lake Huzenbach (Table 1). Most analyzed parameters demonstrate enrichment factors of about 2 to 5 in the canopy throughfall compared to open precipitation. The enrichment factor for manganese reached the number 28, and for potassium 32. The deposition of DOC in canopy throughfall exhibits the absolute greatest value among all measured parameters with about 180 to 190 kg ha^{-1} a. The average Cl/Na ratio in open precipitation was found to be slightly minor than its ratio in sea salt spray. In July 1990, unusual high pH values of 6.5 and 5.6 (canopy throughfall and open precipitation, respectively) were observed in red coloured rainwater samples.

Lake inflows

All the lake inflows are acidic. Inflows mainly draining from the dystric planosol area (e.g. episodic inflow 2) are characterized by extreme discharge oscillations and a fast response of discharge following heavy rains (estimated time-lag: several hours). Inflow 2 is particularly acidic and humic (Table 2). Organic anions exceed inorganic anions applying the model from Oliver *et al.* (1983). Aluminium is on an average 78% organically bound. Alkalinity is always strictly negative. Acidity itself is positively correlated with UV absorption at 254 nm, DOC and SO_4. However,

Table 1. Bulk deposition at Lake Huzenbach.

	Hydr. year	Prec. $l\,m^{-2}$	pH w.m.	H^+ mol ha^{-1}	Ca	Mg	Na	K	Fe	Mn	Al	Cl	NO_3	SO_4	DOC	NH_4-N	DON	TDP
					kg ha^{-1}													
Open	1989	1353	4.42	518	6.7	0.9	4.6	1.5	0.2	0.08	1.0	7.2	24.7	31.0	28.7	5.0	1.2	0.08
Precipitation	1990	1765	4.73	328	6.8	1.1	5.9	2.3	0.2	0.07	1.5	8.9	22.5	28.6	42.3	5.5	1.3	0.10
Canopy	1989	724	3.74	1314	27.5	4.4	9.0	49.1	0.4	2.2	1.3	26.4	81.5	115.5	189.6	5.7	4.4	0.47
Throughfall	1990	1000	3.99	1026	21.7	6.0	14.3	32.1	0.5	2.0	1.2	26.9	80.5	88.0	177.2	6.0	4.6	0.56

DOC – dissolved organic carbon; DON – dissolved organic nitrogen; TDP – total dissolved phosphorus.
w.m. – precipitation volume weighted mean value.
Hydrological years: 1989 (01.11.88–31.10.89); 1990 (01.11.89–31.10.90).

Table 2. Descriptive statistics for selected lake inflows (01.11.1988–31.10.1990).

	Q ($l\ s^{-1}$)	pH	Alk. ($\mu eq\ l^{-1}$)	SO_4 ($mg\ l^{-1}$)	NO_3 ($mg\ l^{-1}$)	DOC ($mg\ l^{-1}$)	Al-t ($\mu g\ l^{-1}$)	Al-i ($\mu g\ l^{-1}$)
Inflow 1								
Min. value	0.2	4.28	− 49	3.88	1.26	1.00	74	0
Max. value	25.5	5.92	+ 54	6.84	3.55	10.80	1040	606
Mean value	2.1	5.13	+ 2	5.03	2.22	3.34	368	146
Std. deviation	3.6	0.49	28	0.70	0.41	2.32	257	168
Inflow 2								
Min. value	0.12 #	3.66	− 228	1.78	0.13	9.10	452	31
Max. value	204.3	4.11	− 76	9.08	1.77	30.70	1111	268
Mean value	25.1	3.90	− 138	5.02	0.89	18.04	728	161
Std. deviation	48.2	0.10	35	1.33	0.44	5.51	126	68

Number of cases: inflow 1: n = 61; inflow 2: n = 27.
Q = discharge; Alk. = Gran alkalinity;
Al-t = total Al, Al-i = inorganic Al (Al-separation after Driscoll, 1984);
real min. value for Q is zero, because inflow 2 is only episodic.

Spearman rank correlation coefficients demonstrate no correlation between discharge of inflow 2 and analyzed parameters like H^+, DOC, SO_4, Al, Fe, K, Ca and Si (Table 3).

Contrawise, a groundwater spring (inflow 1) exhibits less extreme discharge patterns and a slower response of discharge after heavy rains. Its water chemistry is less acidic and less humic (Table 2). Inorganic anions with a major contribution of SO_4 greatly exceed organic anions. Aluminium is on an average 60% organically bound. Mean alkalinity is slightly positive. Spearman rank correlation coefficients demonstrate a distinct control of discharge on important parameters (Table 3). H^+, SO_4, DOC, Fe, and Al are positively correlated with discharge, whereas K, Ca, and Si are negatively correlated with discharge. Acidity itself is positively correlated at a high and similar level with SO_4 and UV absorption at 254 nm, but to a slightly lower rate with DOC.

Average NO_3 values are generally low but subject to rise under stormflow conditions. NO_3 concentrations are commonly higher in spring inflows than in various inflows controlled by subsurface flow (Table 2). This is also valid for peak values

Table 3. Descriptive statistics for selected lake inflows: SPEARMAN rank correlation coefficients (01.11.1988–31.10.1990).

Q	vs.	H^+	DOC	SO_4	Al	Fe	K	Ca	Si
	Inflow 1	0.93	0.62	0.92	0.85	0.62	− 0.94	− 0.70	− 0.76
	Inflow 2	–	–	–	–	–	–	–	–
H^+	vs.	UV254	DOC	SO_4	Al	Fe	K	Ca	SI
	Inflow 1	0.90	0.66	0.91	0.90	0.69	− 0.92	− 0.70	− 0.71
	Inflow 2	0.73	0.67	0.72	0.74	–	–	− 0.48	–

All coefficients are significant at $P < 0.001$.
Number of cases varies from 14 to 61.
Inflow 1 is a groundwater spring, inflow 2 is an episodic tributary.

Table 4. Mean values of Lake Huzenbach surface samples (0 m) (01.11.1988–31.10.1990).

pH	Cond. ($\mu S\ cm^{-1}$)	Alk. ($\mu eq\ l^{-1}$)	DALK ($\mu eq\ l^{-1}$)	UV254 ($l\ m^{-1}$)	DOC	Si	Na	K	Ca	Mg
					($mg\ l^{-1}$)					
4.68	25.2	−26	105	22.1	8.3	1.44	0.63	0.71	1.24	0.29
Cl	NO_3	SO_4	Al-t	Al-o	Fe-t	Fe-o	Mn	TDP	org-N	Chl. *a*
($mg\ l^{-1}$)			($\mu g\ l^{-1}$)							
1.06	0.66	3.85	356	226	111	92	55	13	247	5.8

Number of samples = 17 to 29.
Cond.: Conductivity; Alk.: Alkalinity; Chl. *a*: Chlorophyll '*a*'.
DALK: loss of alkalinity (after WRIGHT), 1983).
TDP: Total Dissolved Phosphorus.
Al-t: total dissolved Al; Fe-t: total dissolved Fe.
Al-o: organically bound Al; Fe-o: organically bound Fe.

during storm events. In some inflows controlled by subsurface flow, NO_3 may be reduced during summer to the detection limit. Comparing the investigation periods 1988/90 and 1985/86 (Table 2; Thies, 1987), higher mean and maximal pH values were registered in those inflows which are mainly drained by groundwater.

Lake

Mean values for lake surface samples are given in Table 4. Phytoplankton blooms by *Chromulina spec.* and *dinoflagellates* (K. Palm, Limnological Institute, University of Freiburg, pers. com.) just below a closed clear ice cover caused high oxygen concentrations (Fig. 1, see 14.01.1990 & 11.02.1990) including supersaturation. This coincided with increased pH values (Fig. 2). After ice break, a fast onset of reducing conditions in the hypolimnion occurs (Fig. 1) which is correlated with rising pH and positive alkalinity values (Figs 2, 3).

Heavy rains with simultaneous snow melt (18.02.1990) and heavy rains in summer

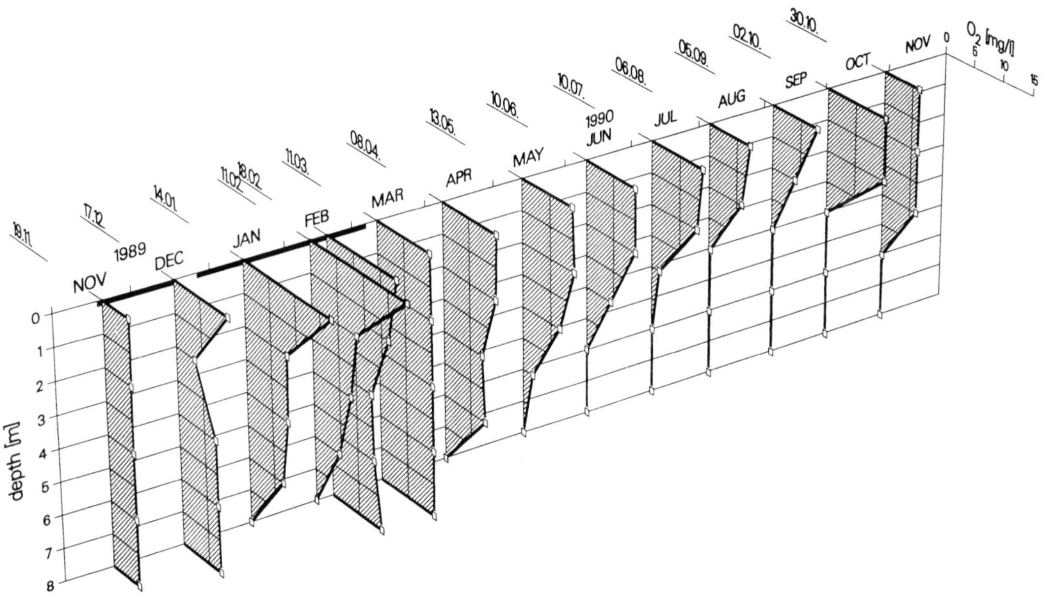

Fig. 1. Dissolved oxygen in vertical profiles of Lake Huzenbach, 1990. Dark bar = ice cover.

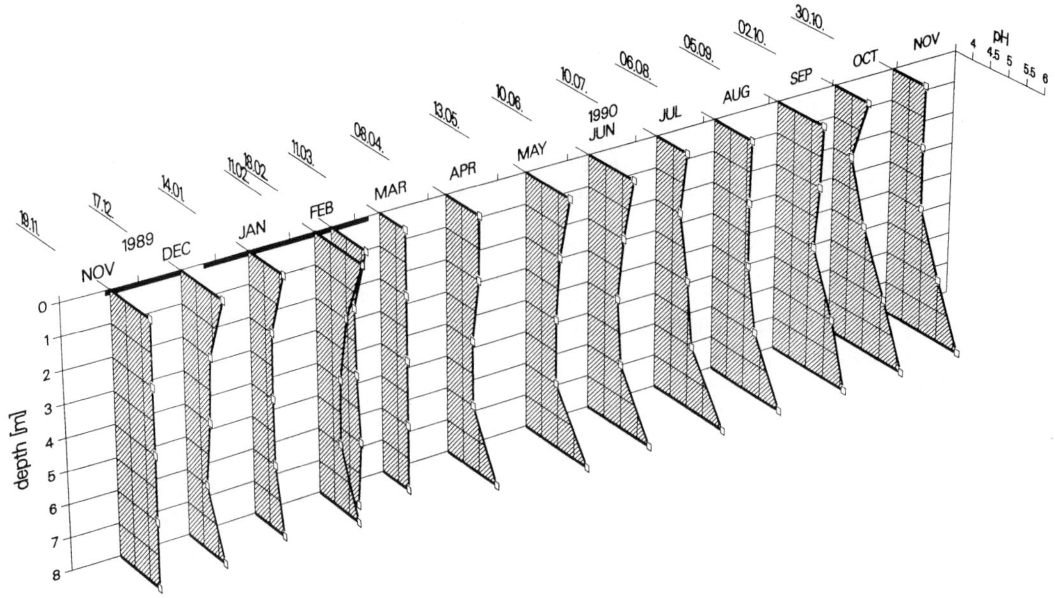

Fig. 2. pH in vertical profiles of Lake Huzenbach, 1990. Dark bar = ice cover.

(10.07.1990 & 02.10.1990) coincided with increasing DOC values in the 2 m stratum of the epilimnion (Fig. 4). Higher DOC values were also found in hypolimnetic layers during anoxic periods. Whereas in summer 1986 no nitrate was detectable in the whole vertical column of the lake for about 4 months (Thies, 1987), this 'nitrate-free' period was found to last only for about 2 months in summer 1990.

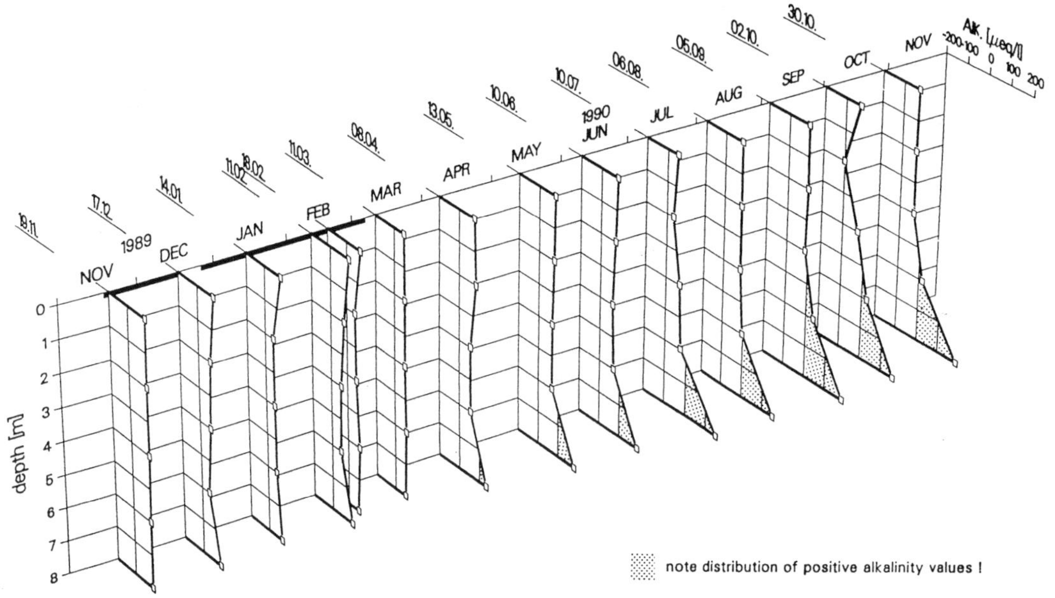

Fig. 3. Akalinity in vertical profiles of Lake Huzenbach, 1990. Dark bar = ice cover.

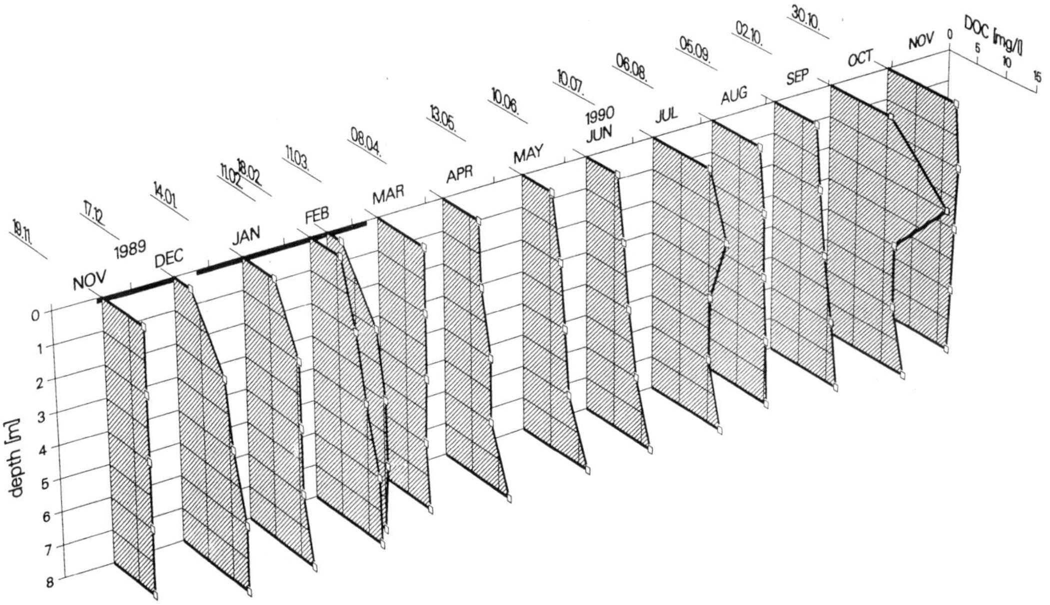

Fig. 4. Dissolved organic carbon in vertical profiles of Lake Huzenbach, 1990. Dark bar = ice cover.

Discussion

Precipitation

Regional trends for a reducing mean annual precipitation amount during the last decades (W. Mikuteit, German Weather Service, Weather Bureau Freiburg, pers. com.) as well as a significant rise in numbers of heavy rains during summertime have been reported recently (von Rudloff, 1991). The rising occurrence of heavy rains is correlated by von Rudloff (1991) with global warming, an enhanced number of condensation nuclei in the troposphere due to air pollution, and changes in distribution patterns of atmospheric pressure. Whether or not the observed heavy rains fits the hypothesis of Rudloff cannot be proved with this short-term study at Lake Huzenbach.

Measured deposition loads at Lake Huzenbach are comparable with those determined by a long-term deposition study by Hochstein & Hildebrand (1991). Calculated enrichment factors (*i.e.* the relation of canopy throughfall to open precipitation) for nitrate and sulfate are greater than three (cf. Table 1). This number is common for conifers receiving very acidic atmospheric pollution (e.g. Adam *et al.*, 1987; Brahmer, 1990; Hochstein & Hildebrand, 1991) and indicates a great impact of dry deposition. The acidic depositions are responsible for the severe forest decline (Hauhs & Wright, 1986) by which conifers are affected. Needle loss, yellowing and subtop-dying are common symptoms of Lake Huzenbach conifers. The order of magnitude for the enrichment factors of potassium and manganese (cf. Table 1) cannot fully be explained by dry deposition and hence indicates an increased leaching from conifer needles (Brahmer, 1990). The destruction of clay minerals in strongly acidified upper soil horizons (Zarei *et al.*, 1991) may have contributed as well to the extreme potassium enrichment factor. It is postulated that ion retention mechanisms within extremely acidified mineral soils do not any longer act as an efficient control for the turnover rates of potassium because of the loss of binding sites with high potassium selectivity (Hildebrand, 1986). Consequently, the circulation velocity for potassium increases between conifer tree foliage and humic top soil layers (Evers *et al.*, 1986) which supports the high enrichment factors.

Although both open precipitation and canopy

througfall are still acidic, precipitation acidity declined from 1989 to 1990 at Lake Huzenbach despite the higher precipitation amount in 1990. This decline in deposition acidity is particularly correlated with a reduction of sulfate deposition in the canopy throughfall (Table 1). Decreasing sulfate depositions have been proved for the period of the last decade at various sites in the northern Black Forest (Adam *et al.*, 1987; Gräf, 1990; Hochstein & Hildebrand, 1991). Declining aerosol sulfate concentrations are reported as well for the southern parts of the Federal Republic of Germany regarding the period 1980–1988 (Schaug, 1990). Nodop (1990) correlates a decrease in atmospheric SO_2 with reduced SO_2 emissions for western Europe with respect to the period 1978–1985. She excludes meteorological variabilities as reasons for the observed reductions in SO_2 using back-trajectory analyses. However, Mylona *et al.* (1990) report that both measurements and calculations of reducing SO_2 concentrations correlate with reported emission reductions in Europe, but meteorological changes may modify this correlation site-specifically.

Nitrate decreased slightly in open precipitation and in canopy throughfall from 1989 to 1990 (Table 1). During the above cited pluriannual studies of Adam *et al.* (1987), Gräf (1990), and Hochstein & Hildebrand (1991), nitrate generally remained at a high deposition level but values demonstrated high standard deviations. However, Mosello *et al.* (1985, 1989) report for northern Italy a strong decline of sulfate but a distinct rise of nitrate in open bulk precipitation without any significant trend of pH regarding the period 1981–1988.

A so-called 'red rain' event with unusual high pH and positive alkalinity values only occured once in July 1990 and coincided with a relatively small rainfall volume. Such an event is caused by long-range transported Saharan dust with high calcium carbonate concentration (Loÿe-Pilot *et al.*, 1986). This phenomenon is considered to be of minor importance regarding the observed rise from 1989 to 1990 of mean annual pH in open precipitation at Lake Huzenbach. However, Loÿe-Pilot *et al.* (1986) reported numerous events of 'red rain' for the island Corsica and calculated a high potential of counteracting rainfall acidity.

DOC in open precipitation at Lake Huzenbach is more related to atmospheric sources of anthropogenic origin than to coniferous needle leaching because of the rather remote position of these rain gauges from forest edges. The white coloured polyethylene funnels of the rain gauges very often exhibited a blackish colour resulting from greasy material which was regularly detected during the cleaning of the funnels. This material contained black particulate matter which was also found on top of nucleopore filters after sample filtration. These particles were identified as carbonaceous particles of different shape and size (E. Hilgers, Limnological Institute, University Freiburg, pers. com.). The same carbonaceous particles, which are known to be derived from the burning of oil (Griffin & Goldberg, 1981), appeared for the first time about 40 years ago within the sediment of Lake Huzenbach (Hilgers *et al.*, 1993).

The rain water samples from open field gauges frequently exhibited a strong smell similar to fuel. It is nuclear whether this derives from the emissions of industries and oil refineries situated in main wind direction west of Lake Huzenbach in the Rhine valley, or from military aircraft regularly using the aerial region for low-level flight training. Emissions caused by dense aircraft circulation (e.g. water vapour, NO_x, hydrocarbons) have so far attracted only little attention despite great possible impact on global climate change (von Rudolff, 1988).

A slightly lower molar Cl/Na ratio in open precipitation than in sea salt might indicate a minor influence regarding emissions of hydrochloric acid from various waste incineration plants which are situated about 40 km west of Lake Huzenbach in the Rhine valley. Rain-out effects at the steep ascent of the mountain-ridges of the westerly part of the northern Black Forest may prevent a further atmospheric transport of hydrochloric acid by the dominating south-westerly winds to the easterly situated Lake Huzenbach. This pattern could be due the high reactivity and hygroscopy of hydrochloric acid in air.

Lake inflows

Correlation analyses (Table 3) support the hypothesis that different patterns regarding the discharge and the chemical composition of the lake inflows are controlled by soil type distribution, soil chemistry (Kolb & Bächle, Institute for Soil Science, University of Hohenheim, pers. com.), and by prevailing flow pathways within the watershed. Hydrologic pathways within planosols are believed to control DOC in adjacent inflows (e.g. inflow 2). McDowell & Wood (1984) reported as well that soil processes control the DOC in stream water. It can be deduced by negative correlation values of discharge with K, Ca and Si concentrations, that silicate weathering is still an effective buffering reaction for a groundwater spring during baseflow conditions. Intensified chemical reactions of slowly percolating groundwater with the solid phase of the bedrock (Bunter sandstone) may result in a stronger weathering of its silicates (*i.e.* potassium feldspars). Positive correlations of DOC and SO_4 with H^+ may indicate, that organic and inorganic acids are both likely to contribute to the total acidity. Gorham *et al.* (1986) and Kerekes *et al.* (1986) demonstrated that acidity in coloured surface waters of Nova Scotia is a function of organic acids derived from peatlands, and strong mineral acidity derived from acidic deposition. The relative contribution of strong mineral acids and organic acids to water acidity was found to differ in various countries (Brakke *et al.*, 1987; Gorham *et al.*, 1986; Kerekes *et al.*, 1986; Kortelainen *et al.*, 1989). However, Brakke *et al.* (1987) concluded, by discrimination of weak and strong acids, that despite a certain contribution of organic acids to acidity, strong mineral acids such as sulfuric acid are the primary reason for recent acidification of humic lakes in southern Norway. Future data evaluation will show to which extent seasonal and site-specific variations are important for inflows of Lake Huzenbach regarding the relative contribution of organic and inorganic acids during differing flow conditions.

Seasonal patterns for NO_3 values in subsurface flow controlled inflows indicate that NO_3 is biologically assimilated within the watershed during the vegetation period despite strong forest decline. However, flash floods caused by heavy rains repeatedly led to a transport of NO_3 from the watershed into the lake inflows and into the lake. Actual observations regarding higher maximum pH values in groundwater inflows are thought to be a consequence of prevailing baseflow which is less acidic. Increasing periods of baseflow throughout the year are resulting from droughts, which will become increasingly important in the future if changes in climatic conditions continue. Reduced sulfur despositions might have had an additional impact on the pH and on the chemical composition of the groundwater, which could be one reason for lower sulfate minima in groundwater inflows during the 1988/90-period compared with the 1985/86-period (Thies, 1993; in prep.). Recent paleolimnological studies on high altitude Austrian lake sediments by Psenner & Schmidt (1992) support the hypothesis given above, that the pH of remote lakes can be controlled both by climatic conditions and by acidic deposition.

Lake

The observed increased pH values under a closed clear ice cover can be related to photosynthetical activity of phytoplankton and to prevailing baseflow conditions (see above). Positive alkalinity and increasing pH values within the anoxic hypolimnion are likely to result from microbial reduction of nitrate and sulfate (Cook *et al.*, 1986). Epilimnetic nitrate values, which are close or below detection limit during summer, are explained by biological uptake of algae (Rudd *et al.*, 1988) and of the floating *Sphagnum* peat mat. Decreasing nitrate concentrations in most lake inflows during summer may also contribute to reduced epilimnetic nitrate values. During the summer 1990 the episodic and subsurface-flow controlled inflow 2 imported turbidity, nitrate and DOC from the watershed into the lake due to two heavy rain events. These major flash floods were apparently responsible for rising values in DOC and nitrate in the upper epilimnion peaking within the 2 m stratum (for DOC see Fig. 4). During the start of the flash flood in July 1990 (05.07.) water temperature of the major inflow (episodic inflow 2) was 13.0 °C and vertical profile temperatures were 15.5 °C (0 m), 15.1 °C (1 m), and 11.8 °C (2 m). During the start of the flash flood in October 1990 (02.10.) water temperature of the major inflow (episodic inflow 2) was 10.3 °C and vertical profile temperatures were 12.7 °C (0 m), 10.6 °C (1 m), and 10.1 °C (2 m). These patterns in water temperatures support the idea that peak values in the 2 m stratum can be explained by interflow (Wetzel, 1983) of the episodic inflow 2, which could additionally be intensified by a draining effect of the lake outlet. The outlet is thought to drain preferably the upper two meters of the epilimnion of both the outer peripheral part and the inner main water body of the lake, which are divided by a floating *Sphagnum* peat mat. The *Sphagnum* peat mat shows near the lake outlet at a water depth of two meters a tubelike junction, by which the inner and the outer part of the lake are connected. Even during base flow divers detected a remarkable water current at this submerse junction. Water currents also exist in the outer peripheral part of the lake which links the inflow 1 (and parts of the episodic inflow 2) directly with the outlet.

Comparing similar discharge periods in different seasons (Fig. 4, 18.02. & 02.10.1990), allochthonous organic compounds seem to vary seasonally with greater importance during late summer (Thies, 1991a), which was reported as well by Eshleman & Hemond (1985).

Conclusion

Lake Huzenbach is a dystrophic mountain lake which was strongly acidified during the latest 150 years by atmospheric deposition (Hilgers *et al.*, 1993). Climatic conditions, acidic deposition and flow pathways of lake tributaries within the watershed are important factors for the actual chemical composition of surface waters. An investigation period of a few years cannot yet prove effects on lake water composition caused by the latest trend for reducing deposition acidity and the expected climate change. Liming of terrestrial areas by helicopter with an amount of about 3000 kg ha^{-1} will be carried out in the northern Black Forest from 1991 on, to cure symptoms of forest decline. The chemical composition of surface waters may be modified (Zoettl & Huettl, 1991), if such a treatment would be applied regularly within watersheds of weakly buffered mountain lakes. Biogeochemical cycles within the watersheds of these humic mountain lakes, including their trophic status and their optical properties, might change. Continuous research using digital recording of pH, conductivity, temperature, and discharge in the spring inflow 1 and the lake outlet may provide early evidence for effects of reducing acidic depositions, liming activities, and climate change on surface waters of Lake Huzenbach.

Acknowledgements

These studies were supported by the Environmental Protection Board of the Federal Republic of Germany (Berlin) [Project 'Wasser 102 04 342'], by the Ministry for Environment of Baden-Württemberg (Stuttgart) [by grant no. 22-88.12 to the Institute of Zoology, University of Hohenheim] and by the Limnological Institute of the Universities Konstanz/Freiburg. The Environmental Protection Agency of Baden-Württemberg (Karlsruhe) generously provided laboratory facilities for further water analyses. B. Rudolph and W. Mikuteit (German Weather Service, Weather Bureau Freiburg) supplied weather data and informations on regional climate change. E. Hochstein and Dr E. E. Hildebrand (Forestry Research Institute Baden-Württemberg, Freiburg) provided data on regional deposition studies. Dr E. E. Hildebrand gave helpful comments on the manuscript. K. Palm (Limnological Institute, University Freiburg) supplied unpublished data on chlorophyll and algae. H. Bächle & E. Kolb (Institute for Soil Science, University of Hohenheim) submitted unpublished data on soil characteristics and soil type distribution within the watershed of Lake Huzenbach.

I owe special thanks to K. Palm and C. Neuner (Limnological Institute, University Freiburg) for the assistive cooperation during the field work and to E. Hoehn (Limnology Bureau Hoehn, Freiburg) for computing assistance. I am grateful to E. May, H. Binder, A. Schneider, U. Gonsior, B. Wand & C. Dröge (Institute of Lake Research, Konstanz & Langenargen) for their support by performance of water analyses. Dr H. Rossknecht and Dr G. Wagner (Institute of Lake Research, Langenargen) gave helpful comments on analytical methods. I am very grateful to my teachers in limnology Prof. Dr J. Schwoerbel (Limnology Institute, University of Konstanz) and Dr R. Schröder (Institute of Lake Research, Konstanz) for their assistance.

References

Adam, K., F. H. Evers & K. Littek, 1987. Ergebnisse niederschlagsanalytischer Untersuchungen in südwestdeutschen Wald-Ökosystemen 1981–1986. Forschungsberichte KfK-PEF 24: 119 pp.

Arzet, K., 1987. Diatomeen als pH-Indikatoren in subrezenten Sedimenten von Weichwasserseen. Diss. Abt. Limnol. Innsbruck 24: 266 pp.

Bächle, H., 1991. Das Bodenmuster um den Huzenbacher See (Nordschwarzwald), Bodenbildung, und seine Bedeutung als Standort. Diplomarbeit am Institut für Bodenkunde und Standortslehre, Univers. Hohenheim, 71 pp. (unpubl.).

Brahmer, G., 1990. Wasser- und Stoffbilanzen bewaldeter Einzugsgebiete im Schwarzwald unter besonderer Berücksichtigung naturräumlicher Ausstattungen und atmogener Einträge. Freiburger Bodenkundl. Abh. 25: 295 pp.

Brakke, D. F., A. Henriksen & S. A. Norton, 1987. The relative importance of acidity sources for humic lakes in Norway. Nature 329: 432–434.

Cook, R. B., C. A. Kelly, D. W. Schindler & M. A. Turner. Mechanisms of hydrogen ion neutralization in an experimentally acidified lake. Limnol. Oceanogr. 31: 134–148.

German Weather Service (ed.), 1992. Deutsches Meteorologisches Jahrbuch, 1989 (in press). German Weather Service, Offenbach a. Main.

German Weather Service (ed.), 1993. Deutsches Meteorologisches Jahrbuch, 1990 (in prep.). German Weather Service, Offenbach a. Main.

DEV 1985. Deutsche Einheitsverfahren zur Untersuchung von Wasser, Abwasser und Schlamm. Verlag Chemie, Weinheim.

Drablos, D. & A. Tollan, 1990. Ecological impact of acid precipitation. Proc. of an int. conf. Sandefjord, SNSF project, 383 pp.

Driscoll, C. T., 1984. A procedure for the fractionation of aqueous aluminium in dilute acidic waters. Intern. J. envir. anal. Chem. 16: 267–284.

Eshleman, K. N. & H. F. Hemond, 1985. The role of organic acids in the acid-base status of surface waters at Bickford watershed, Massachusetts. Water Res. Research 21: 1503–1510.

Evers, F. H., E. E. Hildebrand, G. Kenk & W. L. Kremer, 1986. Boden-, ernährungs- und ertragskundliche Untersuchungen in einem stark geschädigten Fichtenbestand des Buntsandstein-Schwarzwaldes. Mitt. Ver. Forstl. Standortskunde u. Forstpflanzenzüchtg. 32: 72–80.

Frank, M., 1936. Erläuterungen zu Blatt 7216 Gernsbach, Geologische Karte von Baden-Württemberg 1:25.000; Geological Survey Baden-Württemberg, Stuttgart.

Gjessing, E., (ed.) 1991. HUMOR Report no. 1/91. Extended abstracts from Forde-Seminar 1991. NIVA, Oslo, Norway, 64 pp.

Gorham, E., J. K. Underwood, F. B. Martin & J. G. Odgen, 1986. Natural and anthropogenic causes of lake acidification in Nova Scotia. Nature 324: 451–453.

Gräf, E., 1990. Ergebnisse niederschlagsanalytischer Untersuchungen in südwestdeutschen Wald-Ökosystemen. Mitteilungen der Forstlichen Versuchs- und Forschungsanstalt Baden-Württemberg, 151: 99 pp.

Gran, G., 1952. Determination of equivalence point in potentiometric titrations II. Analyst 77: 661–671.

Griffin, J. J. & Goldberg, E. D., 1981. Sphericity as a characteristic of solids from fossil fuel burning in a Lake Michigan sediment. Geochim. Cosmochim. Acta 45: 763–769.

Hauhs, M. & R. F. Wright, 1986. Regional pattern of acid deposition and forest decline along a cross section through Europe. Wat. Air Soil Pollut. 31: 463–474.

Hildebrand, E. E., 1986. Zustand und Entwicklung der Austauschereigenschaften von Mineralböden aus Standorten mit erkrankten Waldbeständen. Forstwiss. Cbl. 105/1: 60–76.

Hilgers, E., Thies, H. & W. Kalbfus, 1993. A lead-210 dated sediment record on heavy metals, polycyclic aromatic hydrocarbons and soot spherules for a dystrophic mountain lake. Ver. Int. Ver. Limnol. 25 (in press).

Hochstein, E. & E. E. Hildebrand, 1991. Stand und Entwicklung der Stoffeinträge in Waldbestände von Baden-Württemberg. Allg. Forst- u. J. -Ztg., 163. Jg., Heft 2: 21–27.

Kerekes, J., S. Beauchamp, R. Tordon & T. Pollock, 1986. Sources of sulfate and acidity in wetlands and lakes in Nova Scotia. Wat. Air Soil Pollut. 31: 207–214.

Kortelainen, P., J. Mannio, M. Forsius, J. Kämäri & M. Verta, 1989. Finnish lake survey: The role of organic and anthropogenic acidity. Water Air Soil Pollut. 31: 235–259.

Loÿe-Pilot, M. D., J. M. Martin & J. Morelli, 1986. Influence of saharan dust on the rain acidity and atmospheric input to the Mediterranean. Nature 321: 427–428.

McDowell, W. H. & T. Wood, 1984. Podzolisation: Soil processes control dissolved organic carbon in stream water. Soil Science 137: 23–32.

Mosello, R., A. Marchetto & G. A. Tartari, 1989. Trends in the chemistry of atmospheric deposition at subalpine and alpine sites in northern Italy. In World meteorological organization (ed.), Changing composition of the troposphere, Special environmental report 17: 33–36.

Mosello, R., G. Tartari & G. A. Tartari, 1985. Chemistry of bulk deposition at Pallanza (N.Italy) during the decade 1975-84. Mem. Ist. Ital. Idrobiol., 43: 311–332.

Mylona, S. N., J. Saltbones, A. Semb & J. Schaug, 1990. Trends in sulfur air quality data in Europe. In Royal society of Edinburg (ed.), Acidic deposition – Its nature and impacts. Abstracts of an international conference, Glasgow, Scotland., 264 pp.

Nodop, K., 1990. How to detect changes in regional emissions by observations. In Norwegian Institute for Air Research (ed.), Monitair 4: 17–21.

Oliver, B. G., E. M. Thurman & R. L. Malcolm, 1983. The contribution of humic substances to the acidity of colored natural waters. Geochim. cosmochim. Acta 47: 2031–2035.

Psenner, R. & R. Schmidt, 1992. Climate-driven pH control of remote alpine lakes and effects of acid deposition. Nature 356: 781–783.

Rudd, J. W. M., C. A. Kelly & D. W. Schindler, 1988. Comment on 'Dynamic Model of In-Lake Alkalinity Generation' by L. A. Baker and P. L. Brezonik. Wat. Res. Research 24: 1825–1827.

Schaug, J., 1990. Trends in aerosol sulfate and sulfar dioxide data. In Norwegian Institute for Air Research (ed.), Monitair 4: 3–5.

Schröder, R., 1991. Relevant parameters to define the trophic state of lakes. Arch. Hydrobiol. 121: 463–472.

Thies, H., 1987. Limnochemische Untersuchungen an vier Karseen des Nordschwarzwaldes unter Berücksichtigung von sauren Niederschlägen sowie der Makrophytenvegetation. Diplomarbeit am Limnologischen Institut der Universität Freiburg, 330 pp. (unpubl.).

Thies, H. & E. Hoehn, 1988. Gewässerversauerung und Limnochemie von sechs Karseen des Nordschwarzwaldes. DVWK Mitteilungen 17: 413–418.

Thies, H., E. Hoehn & R. Schoen, 1988. Gewässerversauerung und Limnochemie von sechs Seen im Nordschwarzwald. Hohenheimer Arbeiten 20: 219–224.

Thies, H., 1990. Acidification studies at northern Black Forest cirque lakes. IAHS Publ. 193: 511–515.

Thies, H., 1991a. Limnochemical studies at small Black Forest lakes (FRG) with special emphasis to acidification. Verh. int. Ver. Limnol. 24: 806–809.

Thies, H., 1991b. Physico-chemical properties of acidified humic Black Forest headwater lakes. In: Gjessing, E., (ed.), 1991. Extended abstracts from Forde-Seminar 1991. HUMOR Report no. 1/91: 30–31. NIVA, Oslo, Norway.

Von Rudloff, H., 1988. Luft- und Raumfahrt als Ozonkiller. Raum & Zeit 37: 22–37.

Von Rudloff, H., 1991. Klimaschwankungen in Europa, Stadteinfluß und Treibhauseffekt. Z. Metorol. 41: 216–226.

Wetzel, R. G., 1983. Limnology. 2nd edition. Saunders College Publishing.

Wright, R. F., 1983. Predicting acidification of North American Lakes. Acid Rain Research Report 4/83: 164 pp., NIVA, Oslo, Norway.

Zarei, M., K. Stahr & K. H. Papenfuß, 1991. Mineralverwitterung und -zerstörung infolge Versauerung in Waldstandorten des Schwarzwaldes. Forschungsberichte KfK-PEF (in press)

Zoettl, H. W. & R. F. Huettl, 1991. Liming as a mitigation tool in declining forests-experiences from former and recent trials. Extended abstract from the 201st meeting of the American Chemical Society, Atlanta, GA, USA, April 14–19, 1991, 183–185.

Hydrobiologia **274**: 155–162, 1994.
J. Fott (ed.), *Limnology of Mountain Lakes.*

Limnological research on northern Apennine lakes (Italy) in relation to eutrophication and acidification risk

Pierluigi Viaroli, Ireneo Ferrari, Gianmarco Paris, Giampaolo Rossetti & Paolo Menozzi
Istituto di Ecologia, Università di Parma, 43100 Parma, Italy

Key words: Northern Apennine lakes, hydrochemistry, phytoplankton, zooplankton, trophic state, acidification

Abstract

Substantial variability was found in the water chemistry of 22 northern Apennine lakes. In a group of lakes there is evidence of disturbance linked to eutrophication processes. Other lakes showed weak ion concentrations and alkalinity below the acidification risk threshold. However no acidified lakes were found. The lack of waterbodies with severely altered hydrochemistry may explain why no clear relationship between plankton community structure and water chemistry was observed.

Introduction

Severe alterations in freshwater chemistry in wide areas of Central and Northern Europe and North America have been linked to acid depositions (Wright, 1983; Wathne *et al.*, 1990). In lakes of these regions a permanent decline in fish populations (Brown, 1987) and considerable changes in structure of benthic (Okland & Okland, 1986) and planktonic (Siegfried *et al.*, 1989; PinelAlloul *et al.*, 1990) communities have been found.

A large research effort has been undertaken to study acidification processes in high altitude lakes in the Alps in relation to acid depositions and watershed lithology (Mosello, 1984; Mosello *et al.*, 1990). The lakes' pristine conditions have been considered particularly suitable for the study of short-term acidification processes even though acidification effects may be masked by other severe environmental conditions (high altitude, extended ice cover period, nutrient limitation) (Psenner & Zapf, 1990).

Little is known about the hydrochemistry and trophic state of lakes located in the peninsular part of Italy, south of the Po river. A regional survey on zooplankton communities of northern Apennine lakes was carried out by Moroni (1962a, 1962b, 1962c). Only recent work reported detailed hydrochemical data (Ferrari, 1976; Antonietti *et al.*, 1987).

A long-term project investigating high altitude northern Apennine lakes was started in 1989. During the first year of research, lithological features were studied and a preliminary survey of hydrochemistry was carried out on 43 lakes from samples collected at the outflow. Wide variability in watershed lithology was found. Several catchments are mostly covered by acidic rocks: lakes located in these areas usually had alkalinity below 0.2 meq l^{-1} and were affected by pulses in pH and ion concentrations due to snow-melt or heavy rainfalls. In spite of this situation, no acidified lakes were found (Viaroli *et al.*, 1992a, 1992b).

Preliminary results of the second year of the project (1990) are reported in this paper. Two main goals were pursued:

- to collect more detailed data on lake water chemistry and trophic state parameters using samples collected on the water column at maximum depth;
- to initiate a long term study of plankton communities and of their potential response to changes in water chemistry.

Materials and methods

In 1990 22 lakes, representative of the original 43, were sampled. The lakes are located on the northern slopes of the western part of the northern Apennines at an altitude between 1000 and 1800 m a.s.l.(maximum altitude is at Mount Cimone, 2265 m a.s.l.) (Fig. 1, Table 1).

Samples from the water column at maximum depth were collected twice during the open water period, in late spring (15 May–15 June 1990) and in fall (26 September–15 October). Temperature and dissolved oxygen were measured at 1 m depth intervals. Water samples for hydrochemistry were collected in all lakes from the surface (−0.5 m) and from the bottom layer using a Ruttner bottle. In lakes 3 to 10 m deep, a sample was also collected at intermediate depth. In the Lake Santo Parmense, whose maximum depth is 22 m, two samples were taken from intermediate layers (−5 and −10 m).

Phytoplankton samples were collected at the same depths also using a Ruttner bottle; an integrated sample for the whole water column was obtained by mixing water from each sampling depth. Quantitative zooplankton samples were collected from the column of maximum depth by vertical hauls with a 50 μm mesh net. Both phytoplankton and zooplankton samples were fixed with buffered formalin.

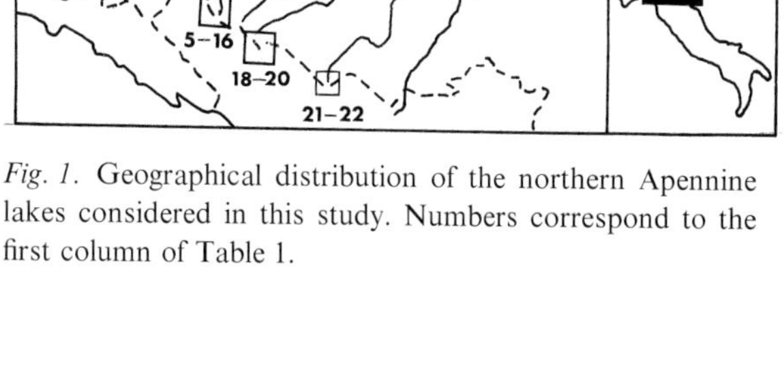

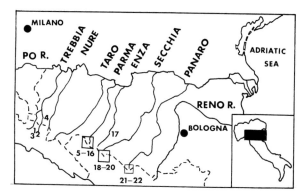

Fig. 1. Geographical distribution of the northern Apennine lakes considered in this study. Numbers correspond to the first column of Table 1.

Water chemistry variables were determined as follows: pH by potentiometry (TIM 90, Radiometer); total alkalinity by potentiometric end-point titration at pH 4.5 and 4.2 (TIM 90, Radiometer) and linearization according to Rodier (1978); conductivity at 20 °C by conductometry (CDM 83, Radiometer); ammonia (Koroleff, 1970), nitrate (Rodier, 1978), chloride and sulphate (see Viaroli *et al.* 1992b), dissolved reactive silica (APHA, 1975), dissolved and particulate phosphorus (Valderrama, 1981) and chlorophyll-*a* (Golterman *et al.*, 1978) by spectrophotometry; calcium, magnesium and sodium by emission spectrometry (Philips PU 7450 ICP); potassium by flame atomic absorption spectrometry (Perkin Elmer 303). Weighted means (by depth) of hydrochemical variables were calculated.

Phytoplankton samples were counted with an inverted microscope; geometric formulae, according to recommendations in Rott (1981), were used to calculate cell volumes. Zooplankton species determination and counting were carried out for rotifers, cladocerans and copepods.

Principal components analysis (PCA) was carried out on logarithm transformed values of conductivity and concentrations of anions, cations, nutrients and chlorophyll-*a*. PCA was also carried out separately on logarithm transformed data of phytoplankton biovolumes and zooplankton densities.

Results and discussion

Lake water chemistry

The mean values of the main hydrochemical variables for the 22 lakes in the spring and fall of 1990 are reported in Tables 2a and 2b.

Most of the lakes have low nutrient and

Table 1. Morphometric and main lithological features of the westernmost lakes in the northern Apennines. H: altitude (m a.s.l.), LA: lake area ($m^2 \times 1000$), WA/LA = watershed area/lake area, Z_{max}: maximum depth (m), ROCK: prevailing surface rock types (MU = maphic and ultramaphic > 50%, SA = sandstone > 89%, SP = sandstone > 89% and peat, ML = marl and limestone and SG = sandstone and gypsum). ND: not determined.

No.	Lake	Valley	H	LA	WA/LA	Z_{max}	ROCK
1	Degli Abeti	Aveto	1329	2.6	166.2	4.6	MU
2	Riondo	Aveto	1328	1.2	200.0	2.5	MU
3	Delle Lame	Aveto	1070	3.6	41.7	7.6	MU
4	Nero	Nure	1540	6.0	41.7	2.5	MU
5	Santo Parmense	Parma	1507	81.6	11.5	22.5	SA
6	Bicchiere	Parma	1724	2.0	27.0	0.9	SA
7	Scuro Parmense	Parma	1527	11.6	18.2	10.4	SA
8	Gemio Superiore	Parma	1355	35.7	47.6	5.6	SA
9	Gemio Inferiore	Parma	1339	32.5	62.4	7.4	SA
10	Pradaccio	Parma	1430	26.9	62.5	3.0	SP
11	Sillara Superiore	Cedra	1732	11.4	6.1	10.6	SA
12	Sillara Inferiore	Cedra	1731	11.4	22.0	10.0	SA
13	Compione Superiore	Cedra	1686	1.8	ND	2.0	SA
14	Compione Inferiore	Cedra	1674	4.7	ND	2.9	SA
15	Verdarolo	Cedra	1390	11.3	38.5	3.0	SA
16	Scuro di Rigoso	Cedra	1392	4.0	76.9	3.0	SA
17	Calamone	Enza	1369	33.5	19.2	9.7	ML
18	Scuro Cerreto	Secchia	1295	2.6	7.6	6.5	SP
19	Cerretano	Secchia	1344	19.2	55.5	4.8	SA
20	Le Gore	Secchia	1337	4.8	92.3	11.0	SG
21	Santo Modenese	Panaro	1501	68.3	13.3	13.7	SA
22	Scaffaiolo	Panaro	1775	11.8	3.4	2.4	SP

chlorophyll-*a* concentrations and may be roughly classified as oligotrophic. In spite of the relatively high altitude, some waterbodies in the Secchia and Enza valleys show symptoms of eutrophication probably due to the effect of tourism pressure.

Maximum total ion concentration is 4.16 meq l^{-1} (Lake Calamone, Enza Valley) with median values (50th percentile) of 0.87 meq l^{-1} in the spring and 0.81 meq l^{-1} in the fall. Both nutrient and ion concentrations show slight seasonal changes. The relationships among the major ions are summarized in Tables 3a and 3b: on the average bicarbonates and sulphates contribute more than 70% of the total anion concentration while calcium and magnesium account for more than 70% of the total cation concentration. In the westernmost valleys (Aveto and Nure), where watersheds are mostly covered by ophiolite rocks, magnesium is more than 50% of the total cation concentration. In the upper Secchia Valley the presence of sulphates among the anions with highest concentration is related to the occurrence of gypsum rocks in the watershed (Viaroli *et al.*, 1992a, 1992b). All lakes located in the most acidic watersheds in the Parma and Cedra valleys (except for the artificially dammed Lake Pradaccio) show ion concentrations below 1.0 meq l^{-1} and conductivities from 20 to 50 $\mu S\ cm^{-1}$: their buffering capacity is weak as shown by alkalinity usually below the critical value (0.2 meq l^{-1}) for acidification risk (Goldstein & Gherini, 1986). However, ion concentrations and their relative importance are within the range reported for alpine lakes (Mosello, 1984; Mosello *et al.*, 1990a).

A synthetic description of lakewater features can be obtained by PCA (Figs 2a and 2b). The first three components explain 72.6% (spring samples) and 73.5% (fall samples) of the total variance. The ordination of the lakes from the first two components seems linked to alkalinity and conductivity (first component) and chloro-

Table 2a. Weighted means (by depth) of trophic state variables determined in spring (S) and fall (F) samples. TP: total phosphorus ($\mu gP\,l^{-1}$), DIN: dissolved inorganic nitrogen (nitrate + nitrite + ammonia, $\mu gN\,l^{-1}$), DRSi: dissolved reactive silica ($mg\,l^{-1}$), Ch-*a*: chlorophyll-*a* ($\mu g\,l^{-1}$). ND: not determined.

No.	Lake	TP		DIN		DRSi		Ch-*a*	
		S	F	S	F	S	F	S	F
1	Degli Abeti	17	22	351	346	2.40	4.34	4.13	0.83
2	Riondo	18	35	180	80	2.22	3.98	3.91	9.36
3	Delle Lame	15	27	142	161	4.25	7.49	2.14	1.59
4	Nero	59	40	156	319	4.32	5.75	13.42	1.12
5	Santo Parmense	8	19	186	103	0.92	0.86	1.61	1.91
6	Bicchiere	30	17	167	437	0.38	0.94	1.47	1.41
7	Scuro Parmense	22	21	210	260	0.46	0.55	0.17	1.59
8	Gemio Superiore	12	24	205	287	0.90	1.01	4.07	1.16
9	Gemio Inferiore	28	20	105	269	0.88	0.77	8.54	2.26
10	Pradaccio	62	17	156	154	0.35	1.05	5.53	0.77
11	Sillara Superiore	12	12	202	166	0.47	0.48	2.24	0.40
12	Sillara Inferiore	14	50	73	24	0.63	0.53	2.51	0.71
13	Compione Superiore	12	18	126	364	1.29	1.27	1.68	0.30
14	Compione Inferiore	9	26	44	86	1.00	1.34	1.10	0.50
15	Verdarolo	21	15	143	196	0.47	1.40	13.20	1.81
16	Scuro di Rigoso	46	17	380	487	0.76	1.38	3.99	5.59
17	Calamone	17	ND	67	ND	0.04	ND	6.46	ND
18	Scuro Cerreto	52	18	102	110	0.78	0.83	39.46	27.85
19	Cerretano	20	17	297	342	2.28	2.42	0.40	5.12
20	Le Gore	27	27	213	171	1.43	1.93	6.79	7.50
21	Santo Modenese	33	29	17	81	0.97	0.85	2.59	1.02
22	Scaffaiolo	38	38	195	305	0.03	0.01	1.02	0.07

phyll-*a* (second component). Accordingly, weakly buffered and eutrophic lakes tend to be grouped in separate clusters.

Plankton communities

Plankton communities of 9 lakes were examined.

Phytoplankton

Biovolumes of phytoplankton samples collected in the spring 1990 (Table 4) seem to distinguish 2 groups of lakes. The first group is formed by three hardwater lakes (Calamone, Le Gore and Santo Modenese) and a soft-water one (Scuro del Cerreto), all affected by anthropic pressure. Phytoplankton biovolume values are greater than $0.3\ mm^3\,l^{-1}$. Lake Scuro del Cerreto showed the highest value ($9.83\ mm^3\,l^{-1}$) primarily due to a large *Cosmarium bioculatum* bloom (more than 80 10^6 cells l^{-1}). A second group of five lakes (Bicchiere, Scuro Parmense, Santo Parmense, Sillara Superiore and Sillara Inferiore), mainly located in the acidic catchments of the Parma and Cedra valleys, has fairly low ($<0.1\ mm^3\,l^{-1}$) biovolume values. For three (Bicchiere, Scuro Parmense and Santo Parmense), small flagellated microalgae ($<10\ \mu m$) belonging to Chrysophyceae and Chlorophyceae make up more than 60% of the total biomass. Phytoplankton of the other two lakes (Sillara Superiore and Sillara Inferiore) are dominated by large forms, mainly Peridiniales and Zygnematales. These differences are reflected by the PCA carried out on biovolume data (Fig. 3); the ordination obtained seems to be more related to trophic state and seasonal succession than to hydrochemical characteristics.

Zooplankton

Spring and fall samples from the same 9 lakes showed relevant differences in number of taxa

Table 2b. Weighted means (by depth) of pH, conductivity at 20 °C (CD, $\mu S\ cm^{-1}$) and major ion concentrations ($meq\ l^{-1}$) in spring (S) and fall (F) samples. Bicarbonates (HCO_3^-) are assumed to be equal to the total alkalinity. ND: not determined.

No.	Lake	Ca^{++}		Mg^{++}		HCO_3^-		$SO_4^=$		pH		CD	
		S	F	S	F	S	F	S	F	S	F	S	F
1	Degli Abeti	0.714	0.520	0.379	0.413	0.951	0.977	0.094	0.096	7.63	7.63	106	104
2	Riondo	0.308	0.236	0.319	0.313	0.509	0.397	0.068	0.139	7.34	6.99	68	55
3	Delle Lame	0.492	0.389	0.790	0.832	1.059	1.007	0.173	0.153	8.29	7.97	115	111
4	Nero	0.048	0.050	0.584	0.676	0.495	0.573	0.074	0.118	7.73	7.43	61	61
5	Santo Parmense	0.250	0.243	0.039	0.044	0.197	0.207	0.089	0.081	7.04	7.03	36	42
6	Bicchiere	0.329	0.374	0.030	0.041	0.277	0.299	0.095	0.095	7.23	7.34	40	50
7	Scuro Parmense	0.154	0.068	0.033	0.037	0.105	0.110	0.091	0.085	6.75	6.68	30	32
8	Gemio Superiore	0.301	0.212	0.041	0.042	0.276	0.141	0.108	0.125	7.21	6.89	38	32
9	Gemio Inferiore	0.186	0.186	0.047	0.047	0.149	0.149	0.075	0.075	6.90	6.90	35	35
10	Pradaccio	0.449	0.338	0.058	0.056	0.406	0.286	0.120	0.148	7.25	7.08	57	51
11	Sillara Superiore	0.082	0.098	0.024	0.030	0.036	0.044	0.065	0.105	6.40	6.56	19	22
12	Sillara Inferiore	0.077	0.085	0.026	0.028	0.042	0.049	0.065	0.060	6.40	6.66	19	24
13	Compione Superiore	0.274	0.270	0.038	0.040	0.253	0.216	0.052	0.076	7.22	7.26	41	42
14	Compione Inferiore	0.219	0.222	0.034	0.039	0.183	0.161	0.074	0.089	7.07	7.16	32	37
15	Verdarolo	0.156	0.161	0.041	0.040	0.101	0.087	0.094	0.097	6.95	6.55	31	33
16	Scuro di Rigoso	0.252	0.189	0.042	0.046	0.162	0.118	0.133	0.091	7.21	6.63	36	38
17	Calamone	1.778	ND	0.316	ND	1.982	ND	0.285	ND	8.37	ND	198	ND
18	Scuro Cerreto	0.072	0.085	0.036	0.040	0.054	0.056	0.065	0.070	6.51	6.24	23	24
19	Cerretano	0.825	0.920	0.118	0.133	0.804	0.934	0.259	0.246	7.87	7.48	97	128
20	Le Gore	1.585	1.651	0.247	0.231	0.972	1.111	0.875	0.870	7.87	7.45	194	214
21	Santo Modenese	0.487	0.609	0.066	0.087	0.529	0.627	0.109	0.133	7.83	7.73	64	81
22	Scaffaiolo	0.589	1.000	0.034	0.050	0.608	0.966	0.067	0.143	7.59	7.89	69	119

and relative abundances of rotifers, cladocerans and copepods. On the whole, 28 species of rotifers, 9 of cladocerans, 6 of cyclopoids and 2 of calanoids were found.

Lake Bicchiere (a small, shallow, high altitude water body) is clearly separated from the other lakes by PCA of both spring (Fig. 4a) and fall (Fig. 4b) density data. In fact its zooplankton are dominated by species which are not found in any of the other lakes, in particular the rotifer *Hexartra mira-intermedia* and the calanoid *Mixodiaptomus tatricus* (calanoids of the other lakes are represented by *Eudiaptomus intermedius*). It might be inferred that the ordination is related to morphometric and physiographic features more than hydrochemical parameters. This conclusion is not surprising and agrees with previous reports by Psenner & Zapf (1990) for Tyrol lakes with an even larger pH range. The PCA of fall samples shows a cluster of three lakes (Le Gore, Santo

Table 3a. Linear regression (y = a + bx) between major anion concentrations (y, $meq\ l^{-1}$) and sum of anion concentrations (x, $meq\ l^{-1}$).

		b	a	r	n
Spring	HCO_3^-	0.774	−0.078	0.965	22
Spring	$HCO_3^- + SO_4^=$	0.984	−0.082	0.999	22
Fall	HCO_3^-	0.725	−0.076	0.958	21
Fall	$HCO_3^- + SO_4^=$	0.993	−0.106	0.999	21

Table 3b. Linear regression (y = a + bx) between major cation concentrations (y, $meq\ l^{-1}$) and sum of cation concentrations (x, $meq\ l^{-1}$).

		b	a	r	n
Spring	Ca^{++}	0.733	−0.055	0.922	22
Spring	$Ca^{++} + Mg^{++}$	0.958	−0.008	0.999	22
Fall	Ca^{++}	0.727	−0.071	0.881	21
Fall	$Ca^{++} + Mg^{++}$	0.982	−0.073	0.999	21

Parmense and Calamone), the only ones in which the microfilter feeders *Bosmina longirostris* and *Ceriodaphnia pulchella* were found. Their presence could be considered a clue of high trophism or of intense predation pressure from planktophagous fish (Kerfoot & Sih, 1987; Richman & Sager, 1990). However this is not confirmed for other lakes in which fish are present and for more clearly eutrophic lakes where small-sized euplanktonic cladocerans were not found.

No clear relationship between zooplankton community characteristics and trophic state or hydrochemistry seems to emerge.

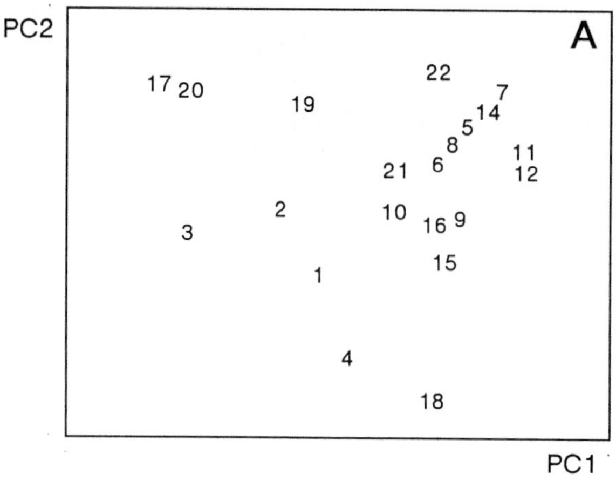

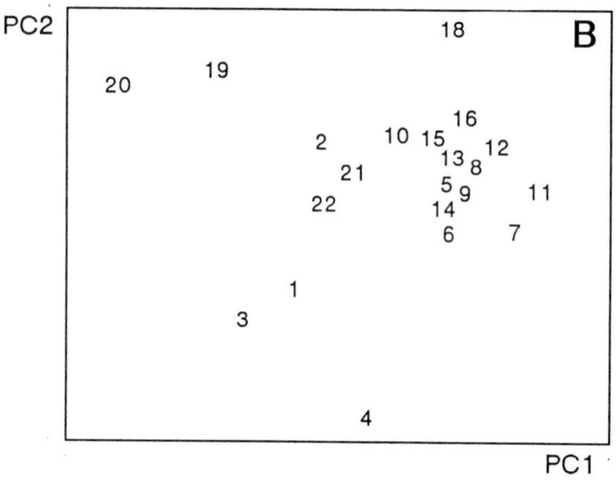

Fig. 2a. Spring 1990: principal components analysis of hydrochemical and trophic state variables (numbers refer to lakes as in Table 1). PC1 (first principal component) and PC2 (second principal component) explain 42.1% and 17.1% of the total variance. PC1 is highly correlated with conductivity ($r = -0.94$) and alkalinity ($r = -0.91$). PC2 is highly correlated with chlorophyll-*a* ($r = -0.76$).

Fig. 2b. Fall 1990: principal components analysis of hydrochemical and trophic state variables (numbers refer to lakes as in Table 1). PC1 (first principal component) and PC2 (second principal component) explain 39.3% and 19.4% of the total variance. PC1 is highly correlated with conductivity ($r = -0.95$) and alkalinity ($r = -0.92$). PC2 is correlated with chlorophyll-*a* ($r = 0.58$).

Table 4. Phytoplankton biovolume values for Spring 1990.

Lake	Biovolume ($mm^3\ l^{-1}$)
Scuro del Cerreto	9.83
Le Gore	1.36
Calamone	1.26
Santo Modenese	0.32
Scuro Parmense	0.09
Sillara Inferiore	0.09
Bicchiere	0.08
Sillara Superiore	0.05
Santo Parmense	0.01

Conclusions

The 1990 study of the water chemistry of 22 northern Apennine lakes, performed on samples

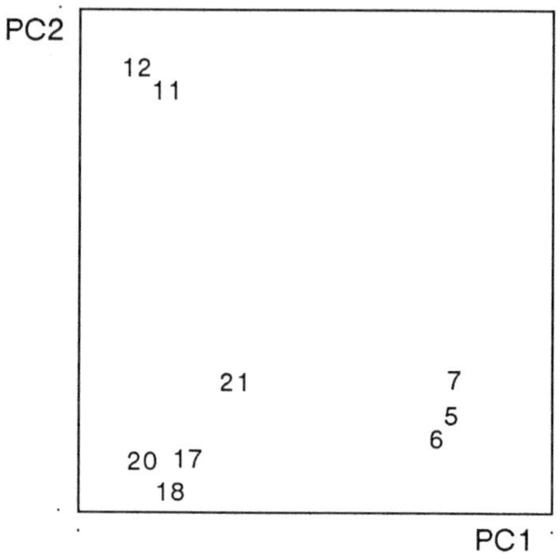

Fig. 3. Spring 1990: principal components analysis of phytoplankton biovolume data (numbers refer to lakes as in Table 1). PC1 (first principal component) and PC2 (second principal component) explain 30.8% and 16.9% of the total variance.

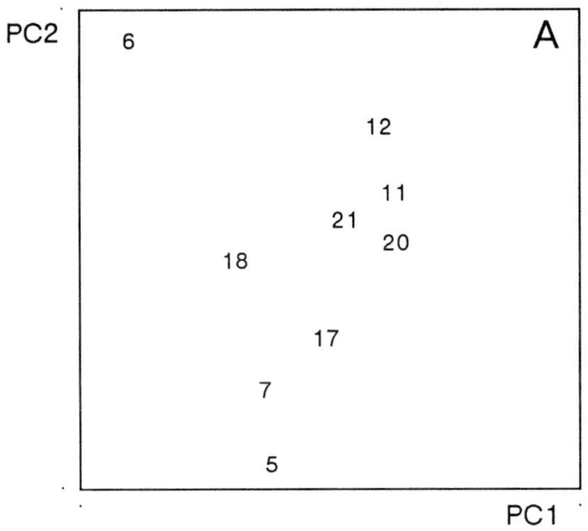

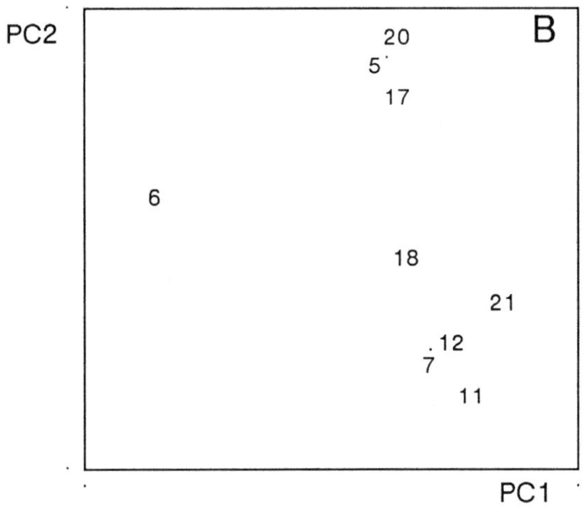

Fig. 4a. Spring 1990: principal components analysis of zooplankton density data (numbers refer to lakes as in Table 1). PC1 (first principal component) and PC2 (second principal component) explain 48.8% and 13.7% of the total variance. *Fig. 4b.* Fall 1990: principal components analysis of zooplankton density data (numbers refer to lakes as in Table 1). PC1 (first principal component) and PC2 (second principal component) explain 49.1% and 12.8% of the total variance.

from the water column at maximum depth, confirmed the results obtained by Viaroli *et al.* (1991a, 1991b) in the previous year on samples collected at the outflow on a larger number of lakes.

In a group of lakes (located in the Enza and Secchia valleys) there is evidence of disturbance linked to eutrophication processes. Other lakes (located in the Parma and Cedra valleys) showed weak ion concentrations and alcalinity below the acidification risk threshold. However no acidified (Henriksen, 1980) lakes were found.

The lack of waterbodies with severely altered hydrochemistry may explain why no clear relationship between phytoplankton (Siegfried *et al.*, 1989) and zooplankton (Psenner & Zapf, 1990) community structure and water chemistry was observed.

Firmer conclusions will probably be reached when larger data sets for long term studies of many lakes will be available.

Acknowledgements

This study has been supported by the CNR-ENEL Project – Interactions of energy systems with human health and environment – Rome, Italy.

References

Antonietti, R., I. Ferrari, G. Rossetti, L. Tarozzi & P. Viaroli, 1987. Zooplankton structure in an oligotrophic mountain lake in Northern Italy. Verh. int. Ver. Limnol. 23: 543–552.

APHA, AWWA, WPCF., 1975. Standard methods for the examination of water and wastewater. 14th edn., APHA, Washington, 1186 pp.

Brown, D. J. A., 1987. Freshwater acidification and fisheries decline. Acid Rain, C.E.G.B. Research 20: 30–38.

Ferrari, I., 1976. Winter limnology of a mountain lake: Lago Santo Parmense (Northern Apennines, Italy). Hydrobiologia 51: 245–257.

Goldstein, R. A. & S. Gherini (eds), 1986. The integrated lake-watershed acidification study. Vol. 4. Summary of major results. Rep. EA-3221 (4). Electric Power Research Institute. Palo Alto, California.

Golterman, H. L., R. S. Clymo & M. A. M. Ohnstad, 1978. Methods for physical and chemical analysis of freshwaters. Blackwell Sci. Publ., Oxford, 213 pp.

Henriksen, A., 1980. Acidification of freshwaters: a large scale titration. Proc. Int. Conf. Ecological impact of acid precipitation In D. Drablos and A. Tollan (eds), Ecological impact of acid precipitation. S.N.S.F. project 1432: 68–74.

Kerfoot, W. C. & A. Sih (eds), 1987. Predation. Direct and indirect impacts on aquatic communities. University Press of New England, Hanover and London, 381 pp.

Koroleff, F., 1970. Direct determination of ammonia in natural waters as indophenol blue. Information on techniques and methods for seawater analysis. I.C.E.S. Interlaboratory rep. No. 3: 19–22.

Moroni, A., 1962a. I laghi della Val Panaro. Boll. Pesca Pisc. Idrobiol., 37 (suppl. 1).

Moroni, A., 1962b. I laghi della Val Parma. Ateneo Parmense, Monografia 8, Parma, 129 pp.

Moroni, A., 1962c. I laghi di Val Secchia. STB, Parma, 80 pp.

Mosello, R., 1984. Hydrochemistry of high altitude alpine lakes. Schweiz. Z. Hydrol. 46: 86–99.

Mosello, R., A. Marchetto, G. A. Tartari, M. Bovio & P. Castello, 1990. Chemistry of Alpine lakes in Aosta Valley (N. Italy) in relation to watershed characteristics and acid deposition. Ambio 20: 7–12.

Okland, D. J. A. & K. A. Okland, 1986. The effects of acid deposition on benthic animals in lake and streams. Experientia 42: 471–486.

PinelAlloul, B., G. Methot, G. Verreault & Y. Vigneault, 1990. Zooplankton species associations in Quebec lakes: variations with abiotic factors, including natural and anthropogenic acidification. Can. J. Fish. aquat. Sci. 47: 110–121.

Psenner, R. & F. Zapf, 1990. High mountain lakes in the Alps: peculiarity and biology. In M. Johannessen, R. Mosello & H. Barth H. (eds), Acidifcation processes in remote mountain lakes. CEC R&D Prog. Air Pollut. Res. Rep. 20: 22–37.

Riehman, S. & P. E. Sager, 1990. Patterns of phytoplankton-zooplankton interaction along a trophic gradient: II. Biomass and size distribution. Verh. int. Ver. Limnol. 24: 401–405.

Rodier, J., 1978. L'analyse de l'eau. Dunod, Orleans, 1136 pp.

Rott, E., 1981. Some results from phytoplankton counting intercalibrations. Schweiz. Z. Hydrol. 43: 34–62.

Siegfried, C. A., J. A. Bloomfield & J. W. Sutherland, 1989. Acidity status and phytoplankton species richness, standing crop and community composition in Adirondack, New York, USA, lakes. Hydrobiologia 175: 13–32.

Valderrama, J. C., 1981. The simultaneous analysis of total nitrogen and total phosphorus in natural waters. Mar. Chem. 10: 109–122.

Viaroli, P., V. Rossi, A. Clerici & P. Menozzi, 1992a. Caratteristiche litologiche ed idrochimiche dei principali laghi dell'Appennino Toseo-Emiliano: risultati preliminari (Maggio–Ottobre 1989). Proc. IX Congr. A.I.O.L., S. Margherita Ligure (Genova, Italy) 20–23 November 1990: 93–104.

Viaroli, P., I. Ferrari, A. Mangia, V. Rossi & P. Menozzi, 1992b. Sensitivity to acidification of Northern Apennines lakes (Italy) in relation to watersheds characteristics and wet depositions. In R. Mosello, B. M. Wathne & G. Giussani (eds), Limnology on groups of remote lakes: ongoing and planned activities. Documenta Ist. ital. Idrobiol. 32: 93–105.

Wathne, B. M., R. Mosello, A. Henriksen & A. Marchetto, 1990. Comparison of the chemical characteristics of mountain lakes in Norway and Italy. In M. Johannessen, R. Mosello & H. Barth H. (eds), Acidification processes in remote mountain lakes. CEC R&D Prog. Air Pollution Res. Rep. 20: 41–58.

Wright, R. F., 1983. Predicting acidification of North American lakes. Acid Rain Res. Rep. 4/83, 165 pp.

Hydrobiologia **274**: 163–170, 1994.
J. Fott (ed.), Limnology of Mountain Lakes.

The effect of anthropogenic acidification on the hydrofauna of the lakes of the West Tatra Mountains (Slovakia)

Marian Vranovský[1], Il'ja Krno[2], Ferdinand Šporka[1] & Jozef Tomajka[1]
[1] *Department of Hydrobiology, Institute of Zoology and Ecosozology, Slovak Academy of Sciences, Drieňová 3, SK-821 02 Bratislava, Slovakia;* [2] *Department of Hydrobiology, Institute of Zoology, Comenius University, Mlynská dolina B 2, SK-842 15 Bratislava, Slovakia*

Key words: zooplankton, macrozoobenthos, anthropogenic acidification, mountain lakes, West Tatra

Abstract

Fourteen West Tatra lakes were studied, of which one could be considered to be recently anthropogenically acidified and eight others classified as acidification-endangered. In the anthropogenically acidified lake, the zooplankton assemblage has been substantially altered (three mountain-lake crustacean species have been eliminated). Several littoral macrobenthic species sensitive to acidification have either been eliminated from the acidified and acidification-endangered lakes or occur only sporadically. The effect of acidification has so far not been observed on the benthic fauna of the lakes medial which is probably due to the higher pH below the surface of sediment. In comparison with the High Tatra, acid depositions have had a less pronounced effect on the lakes of the West Tatra.

Introduction

In contrast to the High Tatra lakes, very little published information is available on the biota and chemical properties of the lakes in the West Tatra (Kubíček & Vlčková, 1954; Obr, 1955; Hrabě, 1961; Štěrba, 1964; Kubíček, 1965; Šporka, 1984; Krčméry *et al.*, 1987). In view of the need to increase our knowledge in this area, the Institute of Fishery Research and Hydrobiology in Bratislava organized a limnological investigation of the West Tatra lakes in 1988–1990. Here we present partial results concerning the hydrofauna and the effect of anthropogenic acidification.

Study sites

The West Tatra and High Tatra are parts of a single mountain complex. The main rock of the both massifs is granite. In the West Tatra there are 20 small lakes of glacial origin, 14 of which were the object of our research. Their basic topographic and morphometric data are shown in Table 1.

Materials and methods

Usually, each lake was sampled twice, once in June and again in September. Water for chemical analyses and physical measurements was collected with the Ruttner sampler from surface and bottom layers. pH was measured by the Radelkis pH-meter, alkalinity by potentiometric titration with 0.01 N HCl with an evaluation according to Gran (Mackereth *et al.*, 1978) and conductivity by the Radelkis conductometer. Ca^{2+} was determined complexometrically and Mg^{2+} by atomic absorption spectrophotometry.

Medial zooplankton were collected with full column vertical tows of a 60–70 μm mesh Apstein net at a uniform speed of approximately

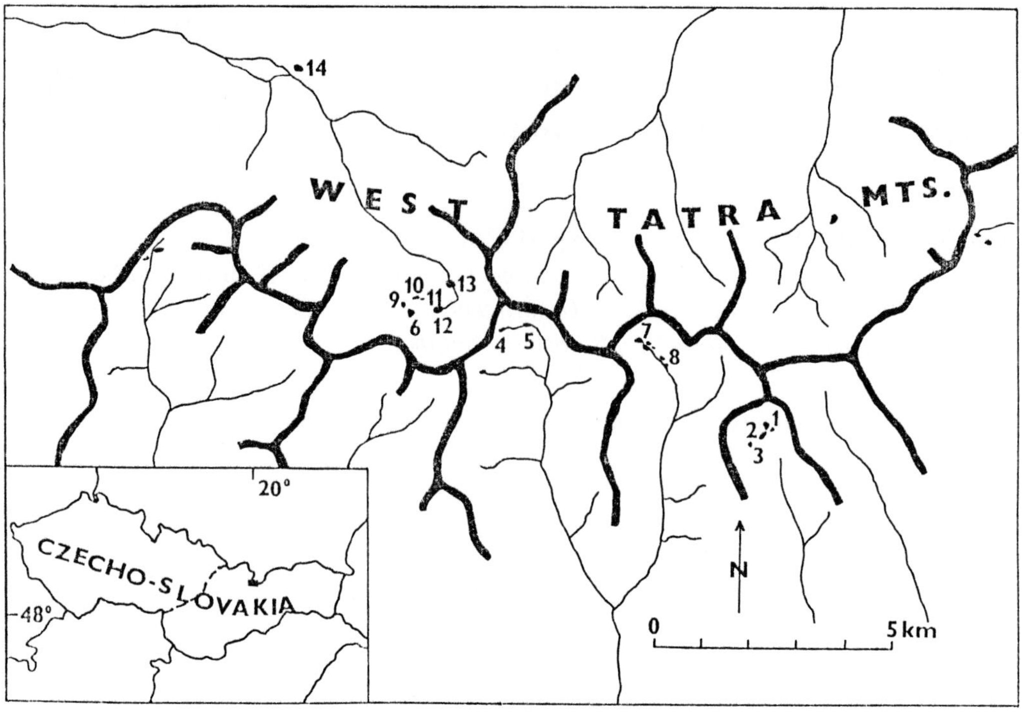

Fig. 1. Map of the West Tatra Mts. The lakes examined are marked with numbers; their names are in Table 1.

20 cm s^{-1}. Littoral zooplankton was sampled with a simple plankton net of the same mesh size. Zooplankton samples were preserved with Lugol's solution and formalin.

Medial macrozoobenthos was sampled with the Ekman-Birge grab and screened with the help of a set of sieves (the densest with mesh openings of 0.5 mm). Littoral macrozoobenthos was

Table 1. Elevation above sea-level and basic morphometric data about the investigated lakes; from Hochmuth *et al.* (1987), some of the values corrected or completed according to our own measurements (x).

No.	Lake	Elevation above sea-level (m)	Area (ha)	Max. depth (m)
1.	Vyšné Bystré "vel'ké" pleso	1876	0.86	12.5
2.	Vyšné Bystré "menšie" pleso	1876	?	5.0ˣ
3.	Nižné Bystré pleso	1837	0.34	4.0
4.	Vyšné Jamnícke pleso	1834	0.41	4.0ˣ
5.	Nižné Jamnícke pleso	1728	1.13	9.0ˣ
6.	Štvrté Roháčske pleso	1718	1.45	8.1
7.	Vyšné Račkovo pleso	1697	0.74	10.0
8.	Malé Račkovo pleso	?	?	3.0ˣ
9.	Opálové pliesko	?	?	?
10.	Tretie Roháčske pleso	1653	0.61	3.7
11.	Druhé Roháčske pleso	1650	0.21	1.3
12.	Prvé Roháčske pleso	1563	2.22	6.5
13.	Ťatliakovo pliesko	1370	0.28	1.2
14.	Zelené pliesko pod Zverovkou	983	0.31	3.7ˣ

Table 2. Mean values of pH, alkalinity (A), conductivity (C_{25}) and Ca^{2+} and Mg^{2+} concentrations in the water of the West Tatra lakes.

Lake	pH	A (μeq l^{-1})	C_{25} (μS)	Ca^{2+} (mg l^{-1})	Mg^{2+} (mg l^{-1})
Vyšné Bystré pleso	6.7	68	42	4.94	1.02
Nižné Bystré pleso	6.7	59	32	3.72	0.30
Vyšné Jamnícke pleso	5.4	8	20	1.34	0.17
Nižné Jamnícke pleso	6.3	29	35	3.00	1.16
Štvrté Roháčske pleso	6.3	35	27	4.69	0.28
Vyšné Račkovo pleso	6.6	50	33	2.69	1.41
Malé Račkovo pleso	6.7	61	35	2.49	1.59
Opálové pliesko	7.1	180	46	7.16	0.35
Tretie Roháčske pleso	6.3	31	22	4.05	0.25
Druhé Roháčske pleso	6.4	44	21	3.94	0.33
Prvé Roháčske pleso	6.7	109	30	5.62	0.58
Ťatliakovo pliesko	7.4	479	67	9.75	2.30
Zelené pliesko pod Zverovkou	4.7	−2	20	1.87	0.25

Table 3. Species composition of zooplankton in the lakes of the West Tatra. Rotatoria.

Taxon / Lake	Štvrté Roháčske pleso	Tretie Roháčske pleso	Druhé Roháčske pleso	Prvé Roháčske pleso	Opálové pliesko	Ťatliakovo pliesko	Zelené pliesko pod Zverovkou	Vyšné Jamnícke pleso	Nižné Jamnícke pleso	Vyšné Račkovo pleso	Malé Račkovo pleso	Vyšné Bystré "veľké" pleso	Vyšné Bystré "menšie" pleso	Nižné Bystré pleso
Rotatoria g. sp.	−	+	+	−	−	+	−	−	−	−	−	−	−	−
Ascomorpha ecaudis (Perty)	−	−	−	+	−	−	−	−	−	−	−	−	−	−
Bdelloidea g. sp.	−	+	+	−	−	−	+	−	−	−	−	−	−	−
Brachionus urceol. f. *sericus* Rousselet	−	−	−	−	−	−	+	−	−	−	−	−	−	−
Cephalodella gibba (Ehrenberg)	−	−	−	−	−	−	−	−	−	−	−	−	+	−
Cephalodella sp.	−	+	−	−	−	−	−	−	−	−	−	−	−	−
Euchlanis dilatata Ehrenberg	−	−	−	+	−	−	−	−	−	−	+	−	+	+
Keratella cochl. f. *cochl.* (Gosse)	−	−	−	−	−	−	+	−	−	−	−	−	−	−
Keratella cochl. f. *tecta* (Gosse)	−	−	−	−	−	−	+	−	−	−	−	−	−	−
Keratella hiemalis Carlin	+	−	−	+	−	−	−	−	−	−	−	+	−	−
Lecane (L.) flexilis (Gosse)	+	+	+	−	−	−	−	+	−	−	−	+	+	−
Lecane (M.) lunaris (Ehrenberg)	−	+	−	−	−	−	+	−	−	+	−	−	−	−
Lepadella patella (O. F. Müller)	+	−	−	−	−	−	−	−	−	−	−	−	+	−
Notholca labis Gosse	+	+	−	−	−	+	−	−	−	−	+	−	−	−
Notholca squamula (O. F. Müller)	−	−	−	−	−	+	−	−	−	−	−	−	−	−
Polyarthra remata Skorikov	−	−	−	+	−	−	−	−	−	−	−	−	−	−
Polyarthra sp.	−	−	−	+	−	−	−	−	−	−	−	−	−	−
Proalidae g. sp.	−	+	−	−	−	−	−	−	−	−	+	−	−	−
Rotaria rotatoria (Pallas)	+	+	−	−	−	−	−	−	−	−	+	−	−	−
Synchaeta tremula (O. F. Müller)	−	−	−	−	−	−	−	−	+	−	−	−	−	−
Trichocerca lophoëssa (Gosse)	−	−	−	−	−	−	−	−	−	−	+	−	−	−
Trichotria tetractis (Ehrenberg)	+	−	−	−	−	−	−	−	−	−	−	−	−	−

Table 4. Species composition of zooplankton in the lakes of the West Tatra. Crustacea.

Taxon / Lake	Štvrté Roháčske pleso	Tretie Roháčske pleso	Druhé Roháčske pleso	Prvé Roháčske pleso	Opálové pliesko	Ťatliakovo pliesko	Zelené pliesko pod Zverovkou	Vyšné Jamnícke pleso	Nižné Jamnícke pleso	Vyšné Račkovo pleso	Malé Račkovo pleso	Vyšné Bystré "veľké" pleso	Vyšné Bystré "menšie" pleso	Nižné Bystré pleso
CLADOCERA														
Acroperus harpae (Baird)	+	+	+	+	–	–	–	+	+	+	–	–	–	–
Alona guttata G. O. Sars	–	–	–	–	–	–	+	–	–	–	–	–	–	–
Alona quadrangularis (O. F. Müller)	–	–	–	–	–	–	–	–	–	–	+	–	–	–
Biapertura affinis (Leydig)	+	+	+	+	–	–	–	+	+	+	–	–	–	–
Ceriodaphnia quadrangula (O. F. Müller)	–	–	–	–	–	–	+	–	–	–	–	–	–	–
Daphnia longispina f. *rosea* G. O. Sars	+	+	+	+	–	–	–	–	+	+	+	–	–	–
Chydorus sphaericus (O. F. Müller)	+	+	+	+	–	+	+	+	+	+	+	+	+	+
Scapholeberis mucronata (O. F. Müller)	–	–	–	–	–	–	+	–	–	–	–	–	–	–
COPEPODA														
Nauplius	+	+	+	+	+	+	+	+	+	+	+	+	+	+
Copepodid (Cyclopoida)	+	+	+	+	+	+	+	+	+	+	+	+	+	+
Acanthocyclops vernalis (Fischer)	+	+	+	+	–	+	+	+	+	–	+	+	–	+
Cyclops abyssorum tatricus Kożmiński	–	–	–	–	–	–	–	–	+	+	+	+	–	–
Eucyclops serrulatus (Fischer)	+	–	+	+	+	+	–	+	+	–	–	+	+	+
Attheyella (A.) wierzejskii (Mrázek)	+	+	–	–	–	–	–	–	–	–	–	–	–	–
Bryocamptus (A.) abnobensis (Kiefer)	–	+	–	–	–	–	–	–	–	–	–	–	–	–
Bryocamptus (A.) cuspidatus (Schmeil)	–	–	–	–	–	–	–	–	+	–	–	–	–	
Bryocamptus (A.) sp.	–	–	–	–	+	–	–	–	–	–	–	–	–	–
OSTRACODA														
Ostracoda g. sp.	+	+	+	–	–	–	–	+	–	+	+	–	–	–
DIPTERA														
Chaoborus obscuripes (V. d. Wulp)	–	–	–	–	–	–	+	–	–	–	–	–	–	–

sampled with the Kubíček sampler (mesh size 0.5 mm). Material was preserved in formalin.

Results and discussion

pH, alkalinity, conductivity, Ca^{2+} and Mg^{2+}

The values of the five, narrowly interconnected parameters presented in Table 2 testify to the fact that only two small lakes (the Ťatliakovo and Opálové) and for the time being the First Roháčske lake too, are not yet immediately endangered by acidification. The lake Zelené pliesko pod Zverovkou had become naturally dystrophic, and the Vyšné Jamnícke lake has been recently anthropogenically acidified. All other lakes are considered to be threatened by acidification (sensu Stuchlík *et al.*, 1985).

Zooplankton

In the majority of the lakes examined, copepods dominated the medial zooplankton (most frequently *Acanthocyclops vernalis* or *Cyclops abyssorum tatricus*). In a smaller number of lakes the cladoceran *Daphnia longispina* f. *rosea* or rotifers (either *Ascomorpha ecaudis*, *Polyarthra remata*, and *Keratella hiemalis* or *K. cochlearis* f. *tecta*) were dominant. In two lakes with subsurface springs as well as in a larger and deeper lake which is already acidified at present (the Vyšné Jamnícke lake), we did not find typical zooplankton assemblages.

Arctodiaptomus alpinus (Imhof) and the high mountain ecotype of *Daphnia pulicaria* Forbes which are well-known from the majority of non-acidified lakes of the subalpine and alpine zones of the High Tatra, are missing in the examined West Tatra lakes and possibly were never established in these lakes.

In most of the lakes for which there exist zooplankton data from the recent past, the species composition found by us was the same or similar to that observed in the 1950s and 1960s (Kubíček & Vlčková, 1954; Kubíček, 1965). However in the dystrophic Zelené lake we did not find *Daphnia longispina* and *D. obtusa*, which were reported by Kubíček & Vlčková (*op. cit.*). Their disappearance may possibly be due to the gradual dystrophication of the lake. Especially noteworthy are the changes in the composition of the crustacean plankton of the Vyšné Jamnícke lake (see Table 5). In contrast to Kubíček & Vlčková (*op. cit.*) we have not found the copepods *Cyclops abyssorum tatricus* and *Mixodiaptomus tatricus* nor the cladoceran *Daphnia longispina* f. *rosea*. Although there are no historical records of alkalinity and pH, there is no doubt that these changes in the zooplankton were caused by anthropogenic acidification. A similar loss of typical mountain-lake crustaceans attributed to acid depositions was reported for numerous High Tatra lakes by Stuchlík *et al.* (1985) and Vranovský (1990). In free water of the Vyšné Jamnícke lake we also identified five species of rotifers not reported by Kubíček & Vlčková. This difference could have been due to the release of the rotifers from the controlling effects of predation after the elimination of *Cyclops abyssorum tatricus*.

Macrozoobenthos of the medial of the lakes

In lakes of the West Tatra, as in High Tatra lakes, oligochaetes and the larvae of chironomids dominated the medial macrozoobenthos. In a few instances Megaloptera (*Sialis lutaria*), Hirudinea (*Erpobdella monostriata*) or Bivalvia (*Pisidium casertanum*) predominated here.

In general, the medial macrozoobenthos of lakes in the High and West Tatras are identical in species composition. One exception is *Erpobdella monostriata* which occurs in two of the Roháčske lakes but has not been found in the High Tatra lakes.

As in the High Tatra lakes, in the West Tatra lakes there is no evidence of acidification affecting the species composition of the medial macrozoobenthos. For example, we found *Pisidium casertanum*, a species of Lamellibranchiata which is sensitive to low pH, in the benthal of the acidified Vyšné Jamnícke lake (pH 5.0 at the bottom in June 1988). In acidified lakes the pH of the upper sediment layers rises with depth (Andersson & Gahnström, 1985). Thus, *P. casertanum*, as a

Table 5. Species composition of zooplankton in the lake Vyšné Jamnícke pleso in September 1951 (Kubíček & Vlčková, 1954) and in September 1988. The frame denotes decline which has been caused obviously by anthropogenic acidification.

Taxon	1951	1988
ROTATORIA		
Lecane (L.) flexilis (Gosse)	−	+
Lecane (M.) lunaris (Ehrenberg)	−	+
Lepadela (L.) patella (O. F. Müller)	−	+
Rotaria rotatoria (Pallas)	−	+
Trichotria tetractis (Ehrenberg)	−	+
CLADOCERA		
Acroperus harpae (Baird)	+	+
Biapertura affinis (Leydig)	+	+
Chydorus sphaericus (O. F. Müller)	+	+
Daphnia longispina f. *rosea* G. O. Sars	+	−
COPEPODA		
Mixodiaptomus tatricus (Wierzejski)	+	−
Cyclops abyssorum tatricus Kożmiński	+	−
Eucyclops serrulatus (Fischer)	−	+

Table 6. Species composition of medial macrozoobenthos in the lakes of the West Tatra.

Taxon / Lake	Vyšné Bystré pleso	Nižné Bystré pleso	Vyšné Jamnícke pleso	Nižné Jamnícke pleso	Štvrté Roháčske pleso	Vyšné Račkovo pleso	Malé Račkovo pleso	Tretie Roháčske pleso	Druhé Roháčske pleso	Prvé Roháčske pleso	Ťatliakovo pliesko	Zelené pliesko pod Zverovkou
TURBELLARIA												
Crenobia alpina (Dana)	–	–	–	–	–	–	–	–	–	+	+	–
OLIGOCHAETA												
Nais variabilis Pig.	–	–	–	+	–	–	–	–	–	–	+	–
Spirosperma ferox (Eis.)	–	–	+	–	–	–	+	–	+	+	–	–
Tubifex tubifex (Müller)	–	–	+	+	+	–	+	+	+	–	+	–
Stylodrilus heringianus Clap.	–	–	–	–	–	+	–	–	+	+	–	
Haplotaxis gordioides (Hart.)	–	–	–	–	–	–	–	–	–	–	+	–
Cognettia sphagnetorum (Vejd.)	–	+	–	–	–	–	–	–	–	–	–	–
Mesenchytraeus sp.	–	–	+	–	–	–	–	–	–	–	–	–
HIRUDINEA												
Erpobdella monostriata (Ged.)	–	–	–	–	–	–	–	+	+	–	–	–
BIVALVIA												
Pisidium casertanum Poli	+	–	+	+	+	–	+	+	–	+	–	–
MEGALOPTERA larvae												
Sialis lutaria (L.)	–	–	–	–	–	–	–	+	+	+	–	–
TRICHOPTERA larvae g. spp.	–	+	–	–	–	–	–	–	–	–	–	–
COLEOPTERA larvae												
Agabus solieri Anb.	–	–	+	–	–	–	–	–	+	–	–	–
Agabus sp.	–	–	–	–	–	–	+	–	–	–	–	–
Helophorus aquaticus (L.)	–	–	–	–	–	–	–	–	–	–	–	+
CHIRONOMIDAE larvae g. spp.	+	+	+	+	+	–	+	+	+	+	+	–
CHAOBORIDAE larvae												
Chaoborus obscuripes (V.D. Wulp)	–	–	–	–	–	–	–	–	–	–	–	+

component of the 'infauna', probably receives greater protection from a decrease in water pH than the benthic 'epifauna', as suggested by Collins *et al.* (1981). However, with continued acidification of the lake-water, the acidity in the sediments will increase and, consequently, the acid-sensitive species of the 'infauna' will be eliminated.

Littoral macrozoobenthos

In the littoral of the West Tatra lakes, the temporary benthic macrofauna (excluding chironomids) is one third as diverse as that found in similar habitats in the High Tatra. In the case of lakes of the Belá river catchment area, one cause of the low species number is the presence of the

Table 7. Species composition of littoral macrozoobenthos in the lakes of the West Tatra.

Taxon \ Lake	Vyšné Bystré pleso	Nižné Bystré pleso	Vyšné Jamnícke pleso	Nižné Jamnícke pleso	Štvrté Roháčske pleso	Vyšné Račkovo pleso	Tretie Roháčske pleso	Druhé Roháčske pleso	Prvé Roháčske pleso	Opálové pliesko	Ťatliakovo pliesko	Zelené pliesko pod Zverovkou
GASTROPODA												
Radix peregra (Müll.)	–	–	–	–	–	–	–	–	+	–	–	–
EPHEMEROPTERA												
Ameletus inopinatus Eat.	–	+	–	–	–	–	–	–	+	–	–	–
ODONATA												
Aeschna juncea (L.)	–	–	–	–	–	–	–	–	–	–	–	+
PLECOPTERA												
Arcynopteryx compacta (McL.)	–	–	–	–	–	–	–	–	–	+	+	–
Capnia vidua Klap.	–	–	–	–	–	–	–	–	–	+	–	–
Diura bicaudata (Kol.)	–	–	–	–	–	–	–	–	+	–	–	–
Leuctra fusca (L.)	–	–	–	–	–	–	–	–	+	–	–	–
Nemurella picteti Klap.	–	–	–	–	+	–	+	+	+	+	+	–
Protonemura auberti Ill.	–	–	–	–	–	–	–	–	–	+	–	–
HETEROPTERA												
Notonecta viridis Del.	–	–	–	–	–	–	–	–	–	–	–	+
MEGALOPTERA												
Sialis lutaria (L.)	–	–	–	–	+	–	+	+	–	–	–	–
COLEOPTERA												
Acilius sulcatus (L.)	–	–	–	–	–	–	–	–	–	–	–	+
Agabus solieri Aub.	–	–	+	–	–	–	+	+	+	–	–	–
Hydroporus incognitus Sharp.	–	–	+	–	–	–	–	+	+	–	+	–
Hydroporus palustris (L.)	–	–	–	–	–	–	–	+	–	–	–	–
Hydroporus planus (Fabr.)	–	–	–	–	–	–	–	–	–	–	–	+
Hygrobius fuscipes (L.)	–	–	–	–	–	–	–	–	–	–	–	+
TRICHOPTERA												
Acrophylax vernalis Dzied.	–	+	–	–	–	–	–	–	–	+	–	–
Drusus monticola McL.	–	–	–	–	–	–	–	–	–	+	–	–
Chaetopteryx fusca Brauer	–	+	–	–	+	–	+	+	+	–	+	–
Limnephilus coenosus Curt.	–	–	+	–	–	–	+	+	–	–	+	–
Oligotricha striata (L.)	–	–	–	–	–	–	–	–	–	–	–	+
Psilopteryx psorosa (Kol.)	–	–	–	–	–	–	–	–	–	+	–	–

fish *Cottus poecilopus* Heckel, which selectively preys on larger forms of may-flies, stone-flies and caddis-flies. In the other lakes the qualitative structure of littoral macrozoobenthos is influenced more by temperature and pH. In the small cold spring-fed lakes (Ťatliakovo and Opálové lakes), there are found species characteristic of springs. In warmer oligotrophic lakes such as the Roháčske lakes (excluding the First) and the already acidified Vyšné Jamnícke lake, species sensitive to acidification, occur only sporadically, although they had been common in the past (Obr, 1955). These taxa include: *Radix peregra*, *Ameletus inopinatus* and *Diura bicaudata*. Dragon-flies, beetles and caddis-flies are abundant in the littoral of the dystrophic Zelené lake. Although several taxa have been eliminated from the West Tatra lakes, the effect of acidification upon the littoral macrozoobenthos has not manifested itself to as great extent as in the High Tatra lakes.

Conclusion

Of the 14 West Tatra lakes studied, one could be considered to be recently acidified while eight others are immediately threatened by anthropogenic acidification.

Acidification has not been uniform in its effects on the components of the hydrofauna.

In the recently acidified Vyšné Jamnícke lake, the zooplankton assemblage has been severely impacted.

Several littoral macrobenthic species sensitive to acidification have either been eliminated from the acidified and acidification-endangered lakes or occur in them only sporadically. The effect of acidification has so far not been observed on the benthic fauna of the medial of the lakes which is probably due to the higher pH below the surface of sediments.

In comparison with the High Tatra, acid depositions have affected a relatively smaller number of lakes in the West Tatra.

Acknowledgements

Mrs E. Hajtmanová, Mrs M. Nagyová and Miss R. Reinoldová provided technical assistance with the laboratory processing and analyses of samples. We also wish to thank Dr Nancy M. Butler for valuable comments and improving the English of the manuscript.

References

Andersson G. & G. Gahnström, 1985. Effects of pH on release and sorption of dissolved substances in sediment-water micro-cosms. Ecol. Bull. 37: 301–318.

Collins, N. C., A. P. Zimmerman & R. Knoechel, 1981. Comparisons of benthic infauna and epifauna biomasses in acidified and nonacidified Ontario lakes. In R. Singer (ed.), Effects of acidic precipitation on benthos. Proc. of a Regional Symp. on Benthic Biol., North amer. benthol. Soc., Hamilton, (N.Y.): 35–48.

Hochmuth, Z. *et al.*, 1987. Západné Tatry. Šport-Slovenské telovýchovné nakladatel'stvo, Bratislava, 281 pp. (in Slovak).

Hrabě, S., 1961. Dva nové druhy *Rhynchelmis* ze Slovenska. Publ. Fac. Sci. Univ. J. E. Purkyně (Brno) 421: 129–146 (in Czech with summaries in Russian and English).

Krčméry, V., K. Bakoš, T. Brezániová, A. Krištinová & A. Krajčír, 1987. Príspevok k hygienickej problematike niektorých prameňov a plies v Západných a východných Tatrách. Geogr. časopis 39: 312–323. (in Slovak with summaries in English and Russian).

Kubíček, F., 1965. Beitrag zur Kenntnis der Art *Cyclops tatricus* Koźmiński (Crustacea, Copepoda) in der Tschechoslowakei. Hydrobiologia 26: 75–84.

Kubíček, F. & D. Vlčková, 1954. Příspevek k poznání zooplanktonu západní jezerní oblasti Liptovských holí. Práce Brněnské základny ČSAV 26, seš. 3, spis 301: 21–48 (in Czech with summaries in German and Russian).

Mackereth, F. J. H., J. Heron & J. F. Talling, 1978. Water analysis: some revised methods for limnologists. FBA Sci. Publ. No. 36: 1–120.

Obr, S., 1955. Příspěvek ke studiu fauny pramenů, jezer a bystřin v Liptovských holích (Tatry). Acta Soc. zool. bohemoslov. 19: 10–26 (in Czech with summaries in Russian and English.

Šporka, F., 1984. Oligochaeta des Flusses Belá. Práce Lab. Rybár. Hydrobiol. 4: 99–117.

Štěrba, O., 1964. Plazivky (Copepoda Harpacticoidea) Moravy a Slovenska. Část I. Acta Univ.Palackianae olomucensis, Fac. Rer. natur. 16: 203–321 (in Czech with summaries in Russian and German).

Stuchlík, E., Z. Stuchlíková, J. Fott, L. Ružička, J. Vrba, 1985. Vliv kyselých srážek na vody na území Tatranského národního parku. Zborník Prác tatr. nár. Parku 26: 173–212 (in Czech with summaries in Russian, German, and English).

Vranovský., 1990. Vplyv antropických faktorov na biocenózy jazier Tatranského národného parku. Zbor. prednášok konf. k 40. výr. uzákonenia Tatran. nár. parku, Tatranská Lomnica: 467–474 (in Slovak).

Hydrobiologia **274**: 171–177, 1994.
J. Fott (ed.), Limnology of Mountain Lakes.

Chlorophyll–phosphorus relationship in acidified lakes of the High Tatra Mountains (Slovakia)

Vojtěch Vyhnálek, Jan Fott* & Jiří Kopáček
*Hydrobiological Institute, Czech Academy of Sciences, Na sádkách 7, 370 05 České Budějovice, Czech Republic; *Department of Hydrobiology, Charles University, Viničná 7, 128 44 Praha 2, Czech Republic*

Key words: chlorophyll, phosphorus, mountain lakes, acidification

Abstract

Concentrations of total phosphorus (TP) and chlorophyll *a* (CHL*A*) were measured in 28 lakes in the High Tatra Mountains (Slovakia) from 1983 to 1990. The relationship between log CHL*A* and log TP in the Tatra lakes is similar to relationships developed for lakes in other regions, but variation is higher. A part of this variation is caused by acidification of the lakes. In the lakes with pH between 4.9 and 6.3 the CHL*A* concentrations are often extremely low while TP concentrations decreased, but not as drastically.

Introduction

The correlation between phytoplankton biomass and phosphorus concentration has been documented for many lakes and reservoirs all over the world. The average summer concentration of chlorophyll is related to the concentration of total phosphorus during the spring overturn (Sakamoto, 1966; Dillon & Rigler, 1974), or to the concentration of total phosphorus during the summer (Jones & Bachmann, 1976; Quirós, 1990).

In this study we summarize results from the investigation of lakes in the High Tatra Mountains (Slovakia) affected by acid rain, obtained during the years 1983 to 1990. The aim is to quantify the chlorophyll – phosphorus relationship in this region and to test the influence of acidification on this relationship.

Material and methods

The High Tatra Mountains (North Slovakia) have been affected by acid rain during recent decades, the annual weighted means of rainwater pH were 4.2–4.5 in this area from 1977–1981 (Cerovský, 1983). The bicarbonate anion in lake water was replaced by sulphate and nitrate during the acidification process, and pH and acid neutralizing capacity (ANC) decreased (Stuchlík *et al.*, 1985).

From a total of 120 lakes in the High Tatra Mountains, 28 were sampled once or twice per year (June/July or September/October) during the years 1983–1990 (Table 1). Summer (June/July) is the time of ice melting, and autumn (September/October) is the end of the vegetation period before the freezing of lakes above timberline (1600–1800 m a.s.l.).

Samples for determination of chlorophyll *a* (CHL*A*) were taken from the surface or were integrated by mixing samples taken at five depths between the surface and the double Secchi depth. Samples for measurement of total phosphorus concentration (TP), chemical oxygen demand (COD), ANC and pH were taken from the surface.

Concentration of CHL*A* was determined in

90% acetone extracts after grinding Whatman GF/C filters with retained seston. Spectrophotometric measurement (Lorenzen, 1967) was used in the years 1983 to 1985, fluorometric procedure (Strickland & Parsons, 1968) was used subsequently. Fluorescence was measured using fluorometer Turner 10 005 R (Turner Designs, USA) or spectrofluorometer SLM (SLM, USA). The fluorometers were calibrated with the spectrophotometric method or with the standard of CHL*A* (Sigma Chem. Co., USA).

Samples for determination of TP and COD were prefiltered through a 40 μm polyamide mesh. Concentration of TP was determined after digestion with perchloric acid (Popovský, 1970) according to Stephens (1963). COD was measured by the bichromate semi-micro method (Hejzlar & Kopáček, 1990).

ANC (Acid Neutralizing Capacity, determined by Gran titration) and pH were measured at room temperature in the laboratory within 24 hrs after sampling.

Simple regressions of log CHL*A* on log TP were calculated in order to enable comparison with similar analyses in literature.

Table 1. Characteristics of 28 lakes studied in the High Tatra Mountains. The lakes are grouped according to pH and ANC. From the lakes denoted (2) or (1), there was at least one sample with pH and ANC within the limits characterizing the groups 2 or 1, respectively. These lakes were classified according to the average values.

Lake	Altitude (m a.s.l.)	Area (ha)	Max. depth (m)	Number of samples
Group 1: pH > 6.3 ANC > 26 μeq l^{-1}				
Štrbské	1346	19.8	20.0	9
Popradské	1494	6.9	17.6	10
Zelené	1545	1.8	4.5	1
Zelené Kačacie	1577	2.6	4.1	1
Velké Biele	1612	0.9	1.0	2
Velické (2)	1663	2.2	4.5	4
Litvorové	1863	1.7	19.1	1
Malé Hincovo	1923	2.2	6.4	7
Ľadové (Zlomisková valley)	1925	2.2	9.6	2
Nižné Terianské	1941	5.5	47.2	1
Velké Hincovo	1946	20.1	53.7	8
Dračie	1961	1.7	16.0	1
Zbojnické (2)	1969	0.6	5.2	2
Pusté	2056	1.2	6.5	2
Ľadové (2) (Velká Studená valley)	2057	1.7	17.8	9
Group 2: pH 4.9–6.3 ANC −7– +26 μeq l^{-1}				
Batizovské	1879	3.5	8.7	8
Velké Žabie (1)	1919	2.6	7.0	2
Dlhé (Velická valley)	1929	0.4	5.0	1
Prostredné Spišské	2013	2.4	4.6	1
Velké Spišské	2014	3.5	10.0	2
Nižné Wahlenbergovo	2053	2.0	8.0	3
Vyšné Wahlenbergovo	2145	5.2	20.0	7
Groups 3: pH < 4.9 ANC < −7 μeq l^{-1}				
Jamské (humic lake)	1447	0.7	4.2	8
Trojrohé (humic lake)	1611	0.3	2.5	2
Slavkovské (humic lake)	1676	0.1	2.5	8
Sesterské	1972	0.3	1.3	2
Starolesnianské	2000	0.7	4.2	6
Vyšné Terianské	2124	0.5	4.2	1

Results and discussion

The lakes under study (Table 1) cover a broad range from ultraoligotrophy to mesotrophy. TP concentrations ranged from the detection limit 0.2 $\mu g\ l^{-1}$ to 21 $\mu g\ l^{-1}$, CHL*A* concentrations ranged between 0.01 and 23 $\mu g\ l^{-1}$. ANC between −50 and 270 $\mu eq\ l^{-1}$ and pH between 4.3 and 7.4 were found during this study.

The relationship between concentrations of CHL*A* and TP (both variables log transformed) is shown in Fig. 1. The linear regressions were calculated for all data, and for summer and autumn data separately. Parameters of our linear regressions (intercept, slope, r^2) are compared with parameters taken from seven other studies relating summer log CHL*A* to log TP (Table 2). Summer and autumn data show significantly different regressions (F-test, $p < 0.05$), but both regressions exhibit nearly the same coefficient of determination (r^2), 0.39 and 0.40, respectively. The difference between summer and autumn data is given mostly by a set of autumn CHL*A* values related to TP that was close to the detection limit.

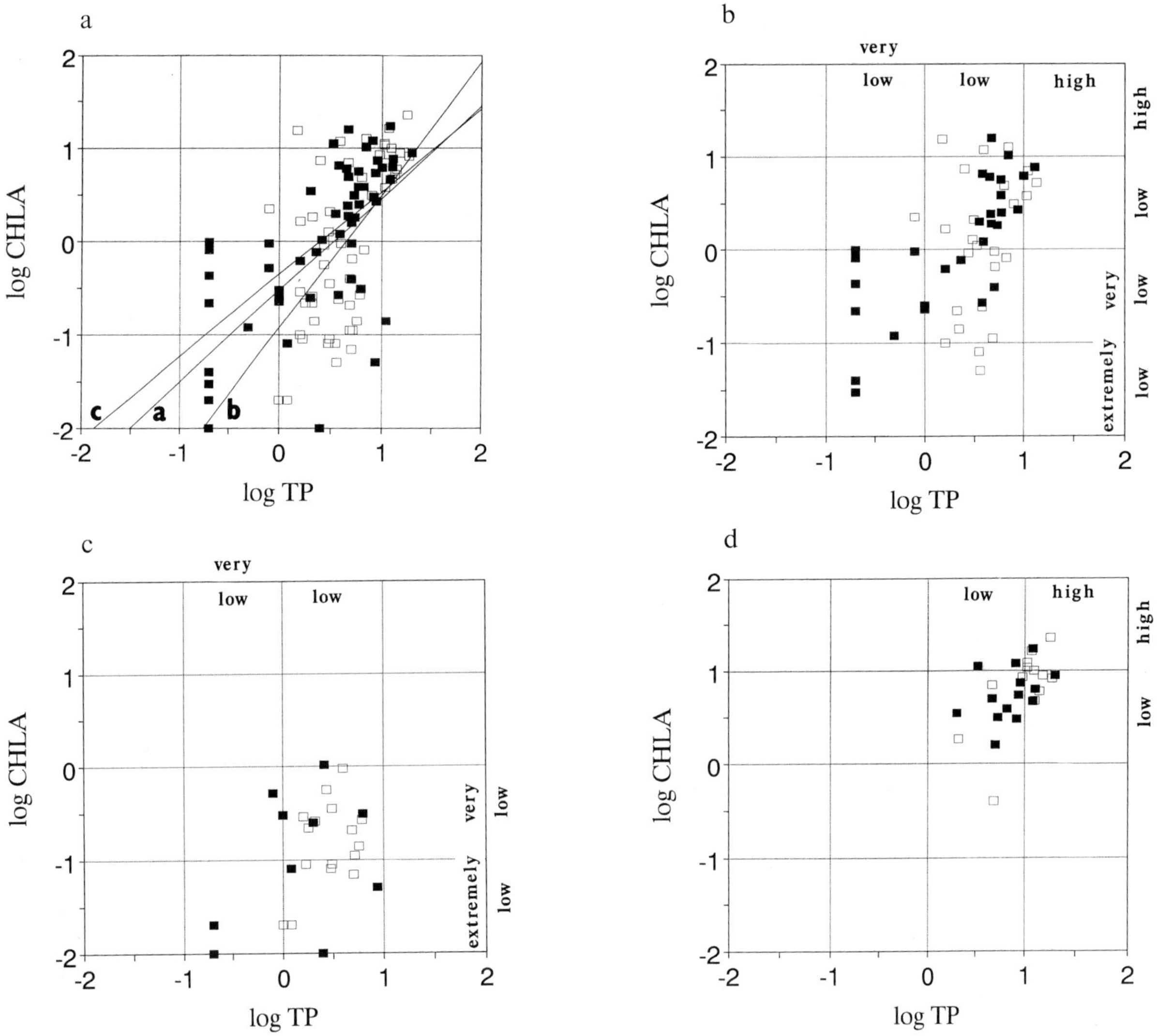

Fig. 1. Relation of chlorophyll (CHL*A*, $\mu g\ l^{-1}$) to total phosphorus (TP, $\mu g\ l^{-1}$) in High Tatra Mountain lakes, both variables log transformed. Open squares – summer data, full squares autumn data. 1a. All data. Line *a* represents linear regression for all data, line *b* for summer data and line *c* for autumn data. Parameters of regressions are given in Table 2. 1b. Data from samples of pH > 6.3 and ANC > 26 $\mu eq\ l^{-1}$: lake group 1. 1c. Data from samples of pH 4.9–6.3 and ANC (−7– + 26 $\mu eq\ l^{-1}$): lake group 2. 1d. Data from samples of pH < 4.9 and ANC < −7 $\mu eq\ l^{-1}$: lake group 3.

These very low TP concentrations might result from sedimentation of phosphorus containing particles (organic and inorganic precipitates, cells and detritus) during the ice-free period.

Parameters of the linear regressions (slopes and intercepts) calculated for High Tatra data are inside the ranges reported in the literature (Table 2). The relationships between log CHL*A* and log TP, however, are remarkably weak in the High Tatra data. Only 35 to 40% of variation in log CHL*A* can be explained by variation in log TP, whereas literature data report much closer relationships ($r^2 > 0.58$) in lakes of different regions (Table 2). Only lakes of the Yukon Territory (Table 2; Shortreed & Stockner, 1986) exhibit a similar weak correlation between the summer log CHL*A* and log TP, probably as a consequence of great density of zooplankton that considerably reduced the phytoplankton biomass.

The following facts could explain the weak cor-

Table 2. Regressions of summer chlorophyll (CHL*A*, $\mu g\, l^{-1}$) on summer total phosphorus (TP, $\mu g\, l^{-1}$) concentrations for different data sets. Both variables are $\log_{10}$ transformed. n = sample size, r^2 = coefficient of determination.

Data	Intercept	Slope	*n*	r^2
This study, all data	− 0.514	0.979	111	0.35
This study, summer data	− 0.922	1.425	55	0.39
This study, autumn data	− 0.344	0.876	56	0.40
Jones & Bachmann, 1976	− 1.090	1.460	143	0.95
Prepas & Trew, 1983	− 0.661	1.146	34	0.81
Stockner & Shortreed, 1985	− 0.090	0.920	50	0.58
Shortreed & Stockner, 1986	− 0.390	0.700	19	0.35
Ostrofsky & Rigler, 1987	− 0.728	1.035	49	0.63
Prairie *et al.*, 1989	− 0.390	0.874	133	0.69
Quirós, 1990	− 1.943	1.080	97	0.78

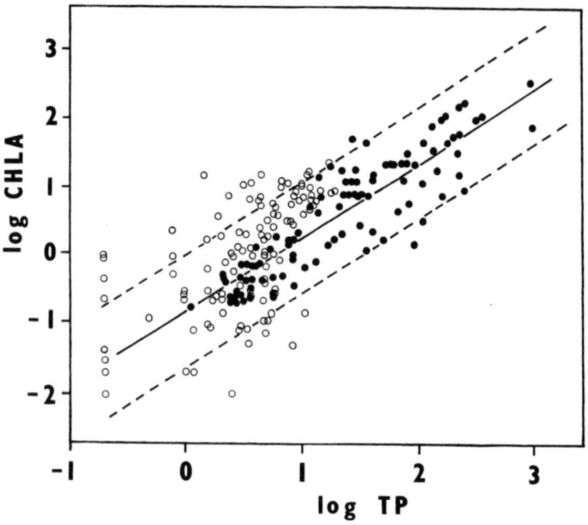

Fig. 2. Relation of summer chlorophyll (CHL*A*, $\mu g\ l^{-1}$) to summer total phosphorus (TP, $\mu g\ l^{-1}$), both variables log transformed. Full circles – data of Quirós (1990) with 95% confidence limits, open circles – data from High Tatra Mountain lakes.

relations between log CHL*A* and log TP in the High Tatra Mountain lakes: 1. discrete samples, not averages, were used in the regressions, 2. a narrow range of TP concentrations, 3. factors other than TP concentration affected the CHL*A* concentration significantly. Only results of Quirós (1990; Table 2) are comparable with ours, because both are based on data from discrete samples. All other results represent average values of TP and CHL*A* concentrations over a summer period. Therefore, we plotted our data into the data set of Quirós, which covers a broad range of TP concentrations including also eutrophic and hypertrophic lakes (Fig. 2). Our data occupy the oligotrophic area of Quirós' range and exhibit higher variation. 31 of 111 values lie out of the 95% confidence interval presented by Quirós. It indicates that higher variation really exists in data from the High Tatra region and that this variation cannot be fully explained just by a narrow range of TP concentrations.

When looking for possible sources of this variation we plotted the log CHL*A* *vs*. log TP relationship separately for three groups of lakes defined according to pH and ANC (Table 1, Fig. 1b, c, d). It appeares that lakes with the lowest pH and ANC (group 3) have both CHL*A* and TP ranging from low to high values. Lakes with intermediate pH and ANC (group 2) have very low or extremely low CHL*A* and low TP.

Another insight may be achieved by plotting CHL*A* and TP concentrations against pH (Fig. 3) and ANC (Fig. 4). Figures 3 and 4 illustrate well the classification of lakes into three groups according to pH and ANC (Table 1).

The three groups of lakes are identical with those defined earlier by Stuchlík *et al.* (1985) and Fott *et al.* (1987, 1994) using an older set of data and slightly different criteria (e.g. lakes below timberline, especially the humic ones, were not included, slightly different pH and ANC limits were used).

Dickson (1980) suggested that an acidified lake turns in the direction to oligotrophy due to precipitation of phosphate, which may occur both in the lake and in the soils of the catchment. According to Jansson *et al.* (1986), the input of phosphorus to the acidified lake Gårdsjön was reduced as phosphate was retained in the soil. Schindler (1988) found no evidence of disruption of the phosphorus cycle in the experimentally acidified ELA lakes.

In the mountain lakes under study, high concentrations of TP were never found in the pH range 4.9–6.3 (Fig. 3), which roughly corresponds to the range of the lowest phosphorus concentra-

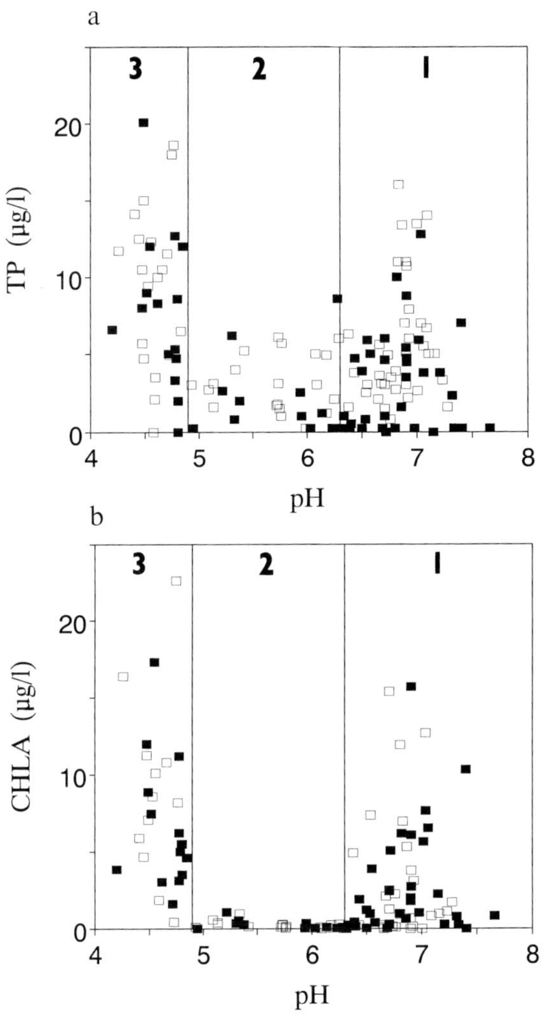

Fig. 3. Relation of total phosphorus (TP – Fig. 3a) and chlorophyll (CHL*A* – Fig. 3b) to pH in High Tatara Mountain lakes. Vertical lines separate data according to pH limits 4.9 and 6.3. Open squares – summer data, full squares – autumn data.

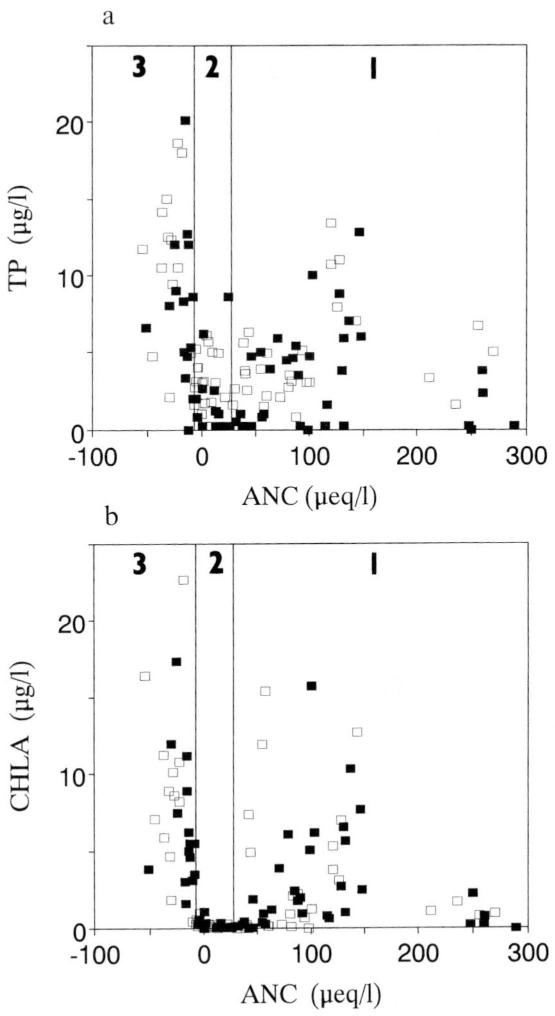

Fig. 4. Relation of total phosphorus (TP – Fig. 4a) and chlorophyll (CHL*A* – Fig. 4b) to acid neutralizing capacity (ANC) in High Tatra Mountain lakes. Vertical lines separate data according to ANC limits −7 and +26 μeq l^{-1}. Open squares – summer data, full squares – autumn data.

tions in Dickson's (1980) laboratory scale experiment. Precipitation of phosphate with metals, especially with aluminium, could be the reason. Precipitation of $AlPO_4$ can be expected at pH 5.5–6.5 (Stumm & Morgan, 1981). An additional possibility, beside the precipitation of phosphate, could be the formation of particulate organic matter containing bound phosphorus. Simultaneous precipitation of organic matter and phosphorus is indicated by similar relationship of TP and COD to pH and ANC (Figs 3, 4 and 6); this aspect was also dealt with by Kopáček & Stuchlík (this volume). The high values of the ratio TP/CHL*A* in the range of pH 4.9–6.3 indicate that this phosphorus was not bound to phytoplankton biomass. Presumably, this phosphorus was present in forms not available to phytoplankton.

The catchments of the lakes are often covered by bare granite rock with little soil. Therefore it is not unreasonable to suppose that precipitation of phosphate may occur in the lake water, but this remains to be tested by studying the phosphorus budgets.

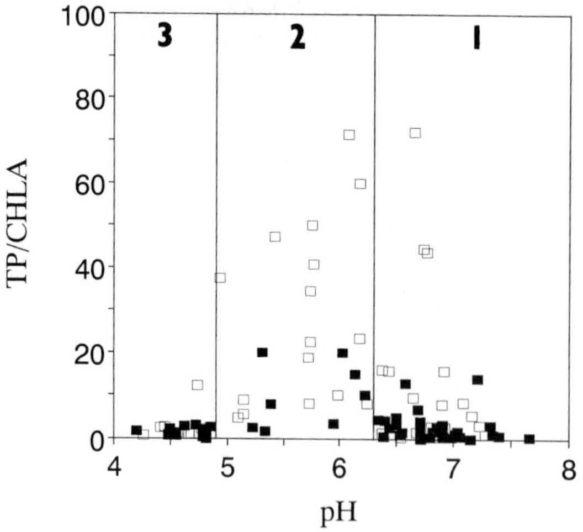

Fig. 5. Relation of total phosphorus:chlorophyll ratio (TP/CHL*A*) to pH in High Tatra Mountain lakes. Vertical lines as in Fig. 3. Open squares – summer data, full squares – autumn data.

Our interpretation of the data on phosphorus, chlorophyll, and pH is that the acidification that shifted pH of the lakes to 6.3–4.9 brought about a further decrease in available phosphorus. This resulted in a decrease of chlorophyll concentrations to very low or extremely low levels. The lakes of group 2 were oligotrophic and P-limited already before the onset of acidification in the early 1970's; the process which had started then may be called ultraoligotrophication. There are no data on chlorophyll and phosphorus from the pre-acidification period, but according to Stuchlík *et al.* (1985) and Fott *et al.* (1987, 1994) the lakes in the pH range 5.0–6.3 lost their zooplankton (Cyclops abyssorum, Arctodiaptomus tatricus), presumably as a result of starvation. If our assumptions are correct, then the disappearance of zooplankton from the acidified high mountain lakes would be another aspect of the shift from oligotrophy to ultraoligotrophy.

The lakes acidified below pH 5.0 are not as deficient in phosphorus and phytoplankton. Zooplankton of these lakes, so far as they are clear-water, non-humic lakes above timberline, are different from those of the pre-acidification time (Stuchlík *et al.*, 1985; Fott *et al.*, 1987, 1994). These lakes are small and shallow.

Kerekes *et al.* (1990) did not find any relation

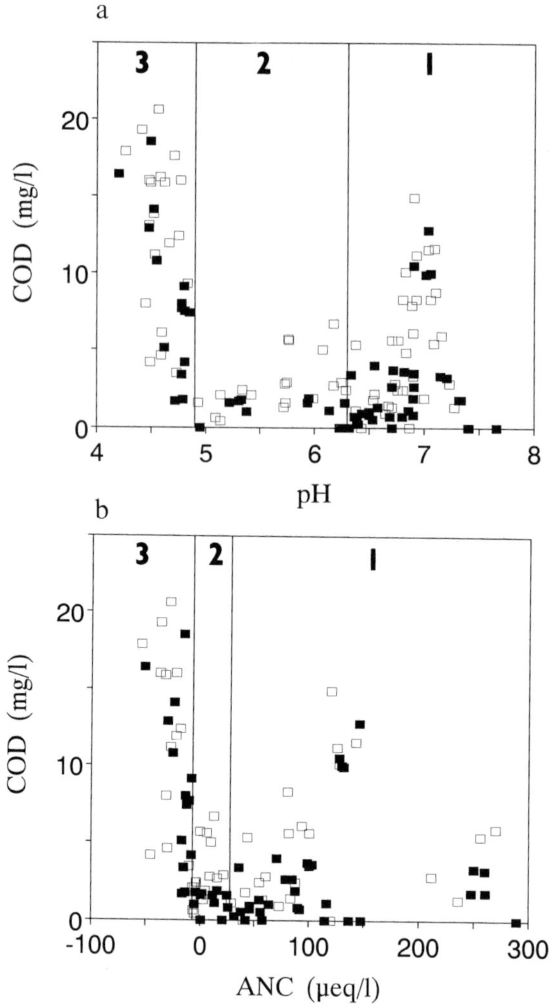

Fig. 6. Relation of COD to pH (Fig. 6a) and COD to ANC (Fig. 6b) in High Tatra Mountain lakes. Vertical lines as in Figs 3 and 4. Open squares – summer data, full squares – autumn data.

of phosphorus and chlorophyll to pH in a group of acidic and non-acidic lakes in Nova Scotia. It may be possible that only oligotrophic, P-limited high mountain lakes are sensitive enough to disrupt their phosphorus cycle in the process of acidification.

References

Cerovský, M., 1983. Measurement of air pollution at Mt. Chopok and pollution of precipitation in Bratislava. (In Slovak). Hydrogeochemical problems of water pollution, Geological Survey, Bratislava: 73–80.

Dickson, W., 1980. Properties of acidified waters. Proc. Int. conf. ecol. impact acid precip., Norway 1980, SNSF project: 75–83.

Dillon, P. J. & F. H. Rigler, 1974. The phosphorus – chlorophyll relationship in lakes. Limnol. Oceanogr. 19: 767–773.

Fott, J., E. Stuchlík & Z. Stuchlíková, 1987. Acidification of lakes in Czechoslovakia. In: Moldan, B. & T. Pačes (eds), Extended abstracts of international workshop on geochemistry and monitoring in representative basins (GEOMON), Prague, Czechoslovakia: 77–79.

Fott, J., M. Pražáková, E. Stuchlík & Z. Stuchlíková, 1994. Acidification of lakes in Šumava (Bohemia) and in the High Tatra Mountains (Slovakia). Hydrobiologia 274/Dev. Hydrobiol. 93: 37–47.

Hejzlar, J. & J. Kopáček, 1990. Determination of low chemical oxygen demand values in water by the dichromate semimicro method. Analyst 115: 1463–1467.

Jansson M., G. Persson & O. Broberg, 1986. Phosphorus in acidified lakes: The example of Lake Gårdsjön, Sweden. Hydrobiologia 139: 81–96.

Jones, J. R. & R. W. Bachmann, 1976. Prediction of phosphorus and chlorophyll levels in lakes. J. Wat. Pollut. Cont. Fed. 48: 2176–2182.

Kerekes, J. J., A. C. Blouin & S. T. Beauchamp, 1990. Trophic response to phosphorus in acidic and non-acidic lakes in Nova Scotia, Canada. Hydrobiologia 191/Dev. Hydrobiol. 53: 105–110.

Kopáček, J. & E. Stuchlík, 1994. Chemical characteristics of lakes in the High Tatra Mountains, Slovakia. Hydrobiologia 274/Dev. Hydrobiol. 93: 49–56.

Lorenzen, C. J., 1967. Determination of chlorophyll and phaeopigments. Spectrophotometric equations. Limnol. Oceanogr. 12: 343–346.

Ostrofsky, M. L. & F. H. Rigler, 1987. Chlorophyll – phosphorus relationships for subarctic lakes in western Canada. Can. J. Fish. aquat. Sci. 44: 775–781.

Popovský, J., 1970. Determination of total phosphorus in fresh waters. Int. Revue ges. Hydrobiol. 55: 435–443.

Prairie, Y. T., C. M. Duarte & J. Kalff, 1989. Unifying nutrient – chlorophyll relationships in lakes. Can. J. Fish. aquat. Sci. 46: 1176–1182.

Prepas, E. E. & D. O. Trew, 1983. Evaluation of the phosphorus – chlorophyll relationship for lakes off the Precambrian Shield in western Canada. Can. J. Fish. aquat. Sci. 40: 27–35.

Quirós, R., 1990. Factors related to variance of residuals in chlorophyll – total phosphorus regressions in lakes and reservoirs of Argentina. Hydrobiologia 200/201/Dev. Hydrobiol. 61: 343–355.

Sakamoto, M., 1966. Primary production by phytoplankton community in some Japanese lakes and its dependence on lake depth. Arch. Hydrobiol. 62: 1–28.

Schindler, D. W., 1988. Experimental studies of chemical stressors on whole lake ecosystems. Verh. int. Ver. Limnol. 23: 11–41.

Shortreed, K. S. & J. G. Stockner, 1986. Trophic status of 19 subarctic lakes in the Yukon Territory. Can. J. Fish. aquat. Sci. 43: 797–805.

Stephens, K., 1963. Determination of low phosphate concentrations in lake and marine waters. Limnol. Oceanogr. 8: 361–362.

Stockner, J. G. & K. S. Shortreed, 1985. Whole-lake fertilization experiments in coastal British Columbia lakes: empirical relationships between nutrient inputs and phytoplankton biomass and production. Can. J. Fish. aquat. Sci. 42: 649–658.

Stuchlík, E., Z. Stuchlíková, J. Fott, L. Růžička & J. Vrba, 1985. Effect of acid precipitation on waters of the TANAP teritory. (In Czech, with English summary). Treatises concerning the Tatra National Park (Bratislava) 26: 173–211.

Strickland, J. D. H. & T. R. Parsons, 1968. A practical handbook of seawater analysis. Bull. Fish. Res. Bd Can. 167, 311 pp.

Stumm, W. & J. J. Morgan, 1981. Aquatic Chemistry, 2nd Edition, Wiley Interscience, New York, 780 pp.

Hydrobiologia **274**: 179–182, 1994.
J. Fott (ed.), Limnology of Mountain Lakes.

Acidification of small mountain lakes in the High Tatra Mountains, Poland

K. Wojtan & J. Galas
Institute of Freshwater Biology, Polish Academy of Sciences, Slawkowska 17, 31-016 Kraków, Poland

Key words: mountain lakes, hydrochemistry, acidification

Abstract

A hydrochemical investigation was carried out on eight small High Tatra mountain lakes (Poland).

When comparing recent data with those from the period 1935–1965, a constant process of acidification of the lakes is found. The average pH of precipitation is 4.8 in the study area, but the lakes are in two stages of acidification: weak (pH 6.0–6.5) and intermediate (pH 5.5–5.8). The differences are due to differences in water sources.

Introduction

Pollutants emitted into the atmosphere from both industry and settlements constitute a major source of degradation of the water environment. Sulphur emissions in Europe have led to the acidification of freshwaters (Wright & Gjeesing, 1976). This type of damage occurs more rapidly in mountain regions with bedrock poor in lime. In Poland, the High Tatra Mountains, whose lakes are extremely susceptible to acidification processes, are such an area (Oleksynowa & Komornicki, 1961, 1989).

This study presents the results of research carried out in the years 1985–1986. Its aim is to determine the degree of acidification of small lakes in the Polish part of the Tatras, taking into account the effect of atmospheric deposition on the chemistry of the lake waters.

Study area

The lakes studied were: Zadni Staw, Dlugi Staw, Czerwony Staw Zachodni and Czerwony Staw Wschodni, Kurtkowiec, Litworowy Staw, Dwoisty Wschodni, and Dwoisty Zachodni (Fig. 1). They are located in the Dolina Gasienicowa valley in the Polish High Tatras, and situated above timberline at altitudes of 1618–1852 m (Table 1). The lakes are of glacial origin, their bedrock is granite, and the buffer capacity of the water is very low (Klimaszewski, 1988; Paschalski, 1963).

Materials and methods

The surface water samples were collected on the following dates: 7 October 1985, 9 June 1986, 18 August 1986, 13 October 1986. From 1986–1987, comparative physical and chemical data were obtained by daily precipitation sampling from the area of the Dolina Gasienicowa valley (from the rain gauge of the Polish Institute of Meteorology and Water Management).

Water and precipitation samples for analysis were taken directly into polyethylene bottles; pH values were measured colorimetrically at the site.

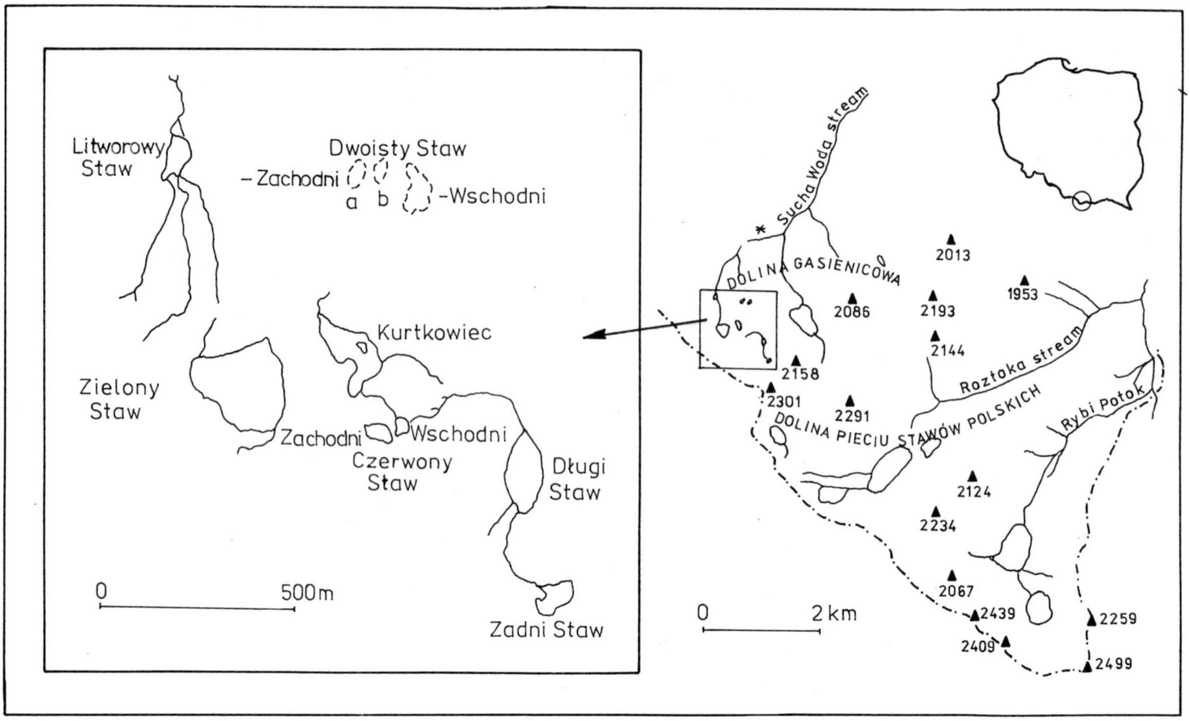

Fig. 1. Map of study area. Dwoisty Staw Zachodni – parts a and b are periodically disconnected on the surface, * – location of the rain gauge.

Commonly accepted methods following Hermanowicz *et al.* (1976) and Standard Methods (1971) were employed in the macrochemical studies. Alkalinity was determined by titration with 0.1 N HCl, and methyl orange as an indicator.

Results and discussion

The mean values of chemical variables in lake water and in precipitation over the study period are presented in Tables 2 and 3.

The precipitation in the Dolina Gasienicowa

Table 1. Morphometric parameters of the lakes under study (according to Klimaszewski 1988).

Name of lake	Altitude (m a.s.l.)	Area (h)	Max. depth (m)	Volume (m^3)
Zadni Staw	1852	0.53	8.0	15340
Długi Staw	1784	1.59	10.6	81060
Czerwony Staw Zachodni	1695	0.27	1.4	1440
Czerwony Staw Wschodni	1693	0.15	1.0	445
Kurtkowiec	1686	1.52	4.8	21200
Dwoisty Zachodni	1657	0.9	7.9	23800
Dwoisty Wschodni	1657	1.41	9.2	48100
Litworowy Staw	1619	0.48	1.1	2720

Table 2. Mean values of chemical variables in the water of 8 small lakes. All units are $\mu eq\ l^{-1}$ except for pH and conductivity ($\mu S\ cm^{-1}$ 18 °C).

Parameter	Litworowy Staw	Kurtkowiec	Czerwony Zachodni	Staw Wschodni	Dwoisty Wschodni	Dwoisty Zachodni	Długi Staw	Zadni Staw
pH	6.2	6.0	6.4	6.5	5.5	5.8	5.6	5.5
Conductivity	21	20	24	29	18	18	19	21
Ca^{2+}	147	140	200	190	110	120	130	140
Mg^{2+}	53	47	58	38	45	39	47	42
NO_3^-	6.8	11.3	10.5	10.7	11.2	12.0	13.2	14.7
PO_4^{-2}	1.7	1.0	1.0	1.2	0.5	0.8	0.8	1.0
SO_4^{-2}	98.4	93.2	97.3	90.9	94.7	85.4	106.2	108.3
Cl^-	31.0	34.4	34.1	33.3	33.5	33.5	33.8	36.9
Alkalinity	130	90	240	140	80	60	60	60
Stage of acidification		Week				Intermediate		

valley was strongly acidified. In 92% of rainfall, the pH was below 5.6 (Wojtan, 1989). The studied lakes, however, are in different stages of acidification. According to the pH and the ratios of $Ca + Mg/SO_4 + NO_3$ values in the lakes, they can be divided into two groups. The first group includes Litworowy Staw, Kurtkowiec, and Czerwony Staw Zachodni and Wschodni, which had pH from 6.0 to 6.4 and the ratios of $Ca + Mg/SO_4 + NO_3$ from 1.8 to 2.2. These lakes are fed by both numerous springs (Wit-Jóżwiakowa & Ziemonska, 1961; Paschalski, 1963) and by water drained through the rock catchment. During this process, weathering takes place, which may increase the buffer capacity of the lake water (Wojtan, unpubl. data).

The second group of lakes, Zadni Staw, Dlugi Staw, Dwoisty Zachodni and Wschodni, had pH below 6, while the ratio of $Ca + Mg/SO_4 + NO_3$ was 1.5. Dlugi Staw and Zadni Staw are highest in altitude at 1800 m, above the dwarf pine level and without water sources other than rainfall and melting snow. The lakes Dwoisty Zachodni and Wschodni, although located at the same altitude as Czerwony Staw Zachodni and Wschodni, have an underground outflow which causes the complete disappearance of water under the snowcover in winter. This group of lakes is fed by a surface inflow and have a higher level of acidification.

Table 3. Precipitation chemistry from 1986–1987. All units are $\mu eq\ l^{-1}$ exept for pH and conductivity ($\mu S\ cm^{-1}$, 18 °C).

Parameter	Range	Mean (volume-weighted)
pH	3.5–6.8	4.8
Conductivity	6–95	32
Ca^{2+}	50–171	78
Mg^{2+}	21–57	40
SO_4^{-2}	2–308	108
Cl^-	28–73	48
Alkalinity	0–80	50

The Tatra area is affected mainly by long-distance emmission from the north (the Upper Silesian industrial region) and from the southwest (Czech and Slovak industrial regions). The additional effects might also have local sources of acid compounds.

A similar investigation on 53 lakes in the Slovak part of the Tatra Mts. found an average pH of 6.1 and ratio of $Ca + Mg/SO_4 + NO_3$ was 1.3. The mean pH in the precipitation was 4.2–4.3 (Fott *et al.*, 1987, 1992).

The SO_4^{-2} ions in the Tatra waters originate mainly from the air, because the crystalline rocks are known to contain only a small amount of sulphates (Oleksynowa, 1970). The mean concentration of sulphates in all studied lakes was about 96 $\mu eq\ l^{-1}$ (Table 2), which is comparable to the mean concentration in precipitation, 108 $\mu eq\ l^{-1}$ (Table 3).

In 1950 there was only a low concentration of chloride (Oleksynowa & Komornicki, 1961) in the surface water, but a subsequent rise to 38 μeq l^{-1} was found 10 years later (Bombówna, 1971). In the present study the chloride concentrations varied from 31 to 37 μeq l^{-1}, while in precipitation the average value was 48 μeq l^{-1}.

The pH values first measured over 50 years ago in 1935 (Olszewski, 1939) were above 6 in all studied lakes. Almost the same values were noted 30 years later (Oleksynowa & Komornicki, 1989) except for Dlugi Staw, where the pH decreased to 5.8. During the next 25 years, a steady fall in pH has been observed in the following three lakes: Zadni Staw, Dwoisty Zachodni, and Wschodni.

A further continuous rise in the concentration of the inflow of acid pollutants can bring about the breakdown of buffering capacity and subsequent acidification, even in those lakes with springs, possibly causing irreversible damage to these ecosystems.

References

Bombówna, M., 1971. The chemical composition of the water of streams of the Polish high Tatra Mts, particularly with regard to the Stream Sucha Woda. Acta Hydrobiol. 13: 379–391.

Fott, J., E. Stuchlík & Z. Stuchlíková, 1987. Acidification of lakes in Czechoslovakia. In Moldan, B., Pačes, T. (eds), Extended abstracts of international workshop on geochemistry and monitoring in representative basins (GEOMON), Prague, Czechoslovakia: 77–79.

Fott, J., E. Stuchlík, Z. Stuchlíková, V. Straškrabová, J. Kopáček & K. Šimek, 1992. Acidification of lakes in Tatra Mountains (Czechoslovakia) and its ecological consequences. Documenta Ist. ital Idrobiol. 32: 69–81.

Hermanowicz, W., W. Dożanska, J. Dojlido & B. Koziorowski, 1976. Physico-chemical examination of water and sewage (in Polish). Arkady, Warszawa, 847 pp.

Klimaszewski, M., 1988. Relief of the Polish Tatra Mts. (in Polish). Panstwowe Wydawnictwo Naukowe, Warszawa, 668 pp.

Oleksynowa, K., 1970. Geochemical characterization of the waters in the Tatra Mountains. Acta Hydrobiol. 12: 1–110.

Oleksynowa, K. & T. Komornicki, 1961. Some new data on the composition of the waters in the Tatra Mountains. Part VI, Valley Rybi Potok and Valley Roztoka. Zeszyt Naukowe Akademii Rolniczej im. H. Kołłątaja w Krakowie. 8: 37–66.

Oleksynowa, K. & T. Komornicki, 1989. Some new data on the composition of the waters in the Tatra Mountains. Part VIII, Valley Sucha Woda. Zeszyty Naukowe Akademii Rolniczej im. H. Kołłątaja w Krakowie. 28: 3–31.

Olszewski, P., 1939. Einige Bestimmungen zum Chemismus der Gewasser in der Umgebung der Gasienicowa-Alm (Hohe Tatra). (in Polish with German summary) Sprawozdanie Komisji Fizjograficznej. 72: 501–530.

Paschalski, J., 1963. An attempt to characterize Tatra waters on the basis of their buffering power. Pol. Arch. Hydrobiol. 11: 349–384.

Standard methods for the examination of water and sewage. 1971 Amer. Public health Association, Washington, 874 pp.

Wit-Jóżwiakowa, K. & Z. Ziemonska, 1962. The hydrography of Polish Tatras. In W. Szafer (ed.), The Tatra National Park. Zaklad Ochrony Przyrody, Kraków: 125–138.

Wojtan, K., 1989. Acid rains in the High Tatra Mts in relation to lakes acidification. In S. Wróbel (ed.), Proc. of the Symp. on Pollution of the atmosphere in relation to the water degradation, (in Polish) Kraków, Poland: 69–75.

Wright, T. R. & E. T. Gjessing, 1976. Acid precipitation: changes in the chemical composition of lakes. Ambio 5: 219–223.